A Boy's Guide
To Big Lake, Minnesota
and
Other Stuff

By
Rod Hunt

Published in the United States by Çiara Press
Postal Box 103, Big Lake, MN 55309 USA
Tel 763 263 5002

Library of Congress Control Number application
governing copyright
submitted September, 2002

ISBN 0-9712693-2-7

SAN 254-1505

Printed in the United States of America

Table of Contents
A Boy's Guide to Big Lake, Minnesota
and
Other Stuff

Dedication iv

Introduction .... v - vi

Acknowledgement vii

Part I... viii

The Big Lake Amusement Park and Dance Pavilion
and Picnic Grounds 1

Nickel or No Nickel 13

What to Do With Rivers 23

How to Make Money So You Can Go to August
Peterson's Drug Store and Ice Cream Parlor in Big
Lake, Minnesota and Enjoy Free Movies 33

Puppies and Kittens and Cats and Dogs 39

How to Become a World Famous Basketball Player in
Big Lake, Minnesota 45

Brown's Hotel in Big Lake, Minnesota 51

Good Eats 59

Bag Swings 67

How to Play Pool 73

How to Play Tackle Football 81

How to Feed Rabbits, Ducks, Pigs and Fences 89

How to Learn Music 99

Huntington's Pasture 109

Tree Houses, Dug-Outs and Lean-Twos 119

How to Go Camping 131

How to Get Ready For Christmas 145

How to Mow Your Lawn 159

How to Get Ready for School in Winter 167

Red Oak 175

How to Spend Your Nights During Winter in Big
Lake, Minnesota 181

Uncle Doc 187

Going to Your Uncle Oscars 195

Pickles, Sauerkraut and Other Stuff 209

Root Beer 215

How to Watch Your Mother and Father 219

Part II 233

Poetry of Sorts 234

    In The Evening 235
    In The Craw of a Crow 236
    The Death Cycle 240
    Up From Down 241
    Dignity 242
    Invitation to a Former Lover 243
    Old Shadows 244
    Hear O Isreal! 245
    Ireland 247

Shortwave at Night.. 248

Local Hero 251

Adam's Apple, A Gift for a Teacher 265

OTZ, AK 289

May Day in September 309

The Dead Do Stir.. 325

DEDICATION

/A little song,
A little dance,
A little seltzer
Down your pants/
........W. C. Fields, among others

This book is dedicated to the older people of Big Lake, Minnesota and their friends in the earnest and sincere hope that it brings back some memories and reminds them of those often lauded "good old days." That would have been when a nickel was a nickel and enough for a good cigar, available, along with ice cream, root beer and good times at August Peterson's Drug Store, and penny candy was in the window of Garver's Store every day of the year and for every holiday or occasion.

The second part of the book, a collection of stories and poems, is dedicated to all the brave souls who can make sense of them or who find meaning and enjoyment in them. That should cover just about everyone.

.....Rod Hunt

Introduction

Like the fella says, "I've had to live without much money and I've had the good fortune to make a little money, and believe me, living with some money is a lot better."

That's sort of like Big Lake. I've lived in Big Lake and I've lived in a lot of other places. Believe me, living in Big Lake is better than a lot of places I could mention.

The Big Lake I have written about is through the eyes of the nine year old I once was. I have some idea of what a young boy who lives here now has. There are certainly some advantages, but I must confess, in the days when we had the good fortune to be able to do things for ourselves, it was certainly a more challenging and trusting time. If we were ever bored, we surely weren't aware of it. That was witnessed in all the things we did on our own. A lot of kids probably do somewhat similar things, but there seems to always be the rather all-encompassing hand or hands of the parents or some supervisor. We didn't spend much time discussing our creative needs or talents. We just did it. I think we must have been pre-indoctrinated with what later became the philosophy of that lady Admiral of the U. S. Navy who proclaimed back in the late 1980's that it was a lot easier to apologize than to get permission.

If Part Two; the collection of short stories and poems, has any purpose other than to entertain you, it is to give pause to the direction we seem to be headed, or others are headed. Some is blatantly political and social, and is somewhat reminiscent of the days of protest a decade or more back in time. All were written critically. Some are specific and some are abject generalizations, which was the nature of the particular beast I was riding at that moment in time. They were written from personal experience, which is the way I like it to be, but you may be able to find agreement with a part or parts. They were written over a wide range of years and in the variety of global locations I happened to be when the muse spoke.

An old friend from near here, after reading some of the stuff a couple of years ago, came up with an opinion about the contents that's stuck with me because he meant no harm and I liked it. He said, "I've been around the world three times and two county fairs in Texas and I've never read anything like it!"

That's plenty good enough for me.
Enjoy!

.....Rod Hunt, Big Lake, Minnesota
September, 2002

Acknowledgment
&
Credits

I would like to thank my wife, Norma Kolbinger, for her help and encouragement in putting these stories back together; my niece, Joyce Ellickson for her critical eye and helpful comments in proofing the whole thing, and Brenda Hill, the indomitable typesetter and good friend.

It should also be noted that Jim Allen, of Allen Studios in Park Rapids, Minnesota did all the illustrations for Part I and the drawing in Part II, for May Day is September is by Anne Lockridge. The drawing of the Holstein doing the Levade Leap for Local Hero is by the above credited Norma.

The local photographs interspersed with the stories are from the Bill and Marg Moores' collection, the Preferred Bank of Big lake and reproduced by Photo One.

The front cover and back cover are from those resurrected by The Big Lake Historial Society.

PART ONE

Part One consists of stories
which will lead to a better life.

Some were previously published
in The West Sherburne Tribune
and the History of Big Lake

Disclaimer
Copyrights of the four numbered photos of Big Lake,
Minnesota appearing herein cannot be located or con-
firmed to be in the Library of Congress Register. Efforts to
locate and contact Mr. Larry Blom of Photo 'N Wood to
whom the photos are credited for reproduction permission
have been unsuccessful despite efforts through the last
available address of 2943 Chowen Avenue North,
Robbinsdale, MN 55422, tel: 612 588 9217 or in the new
area code of (763). Further attempts through Connections,
Etc. Worldwide Service Assistance, www.google.com., and
www.switchboard.com have also proved unsuccessful.

The Big Lake, Minnesota Amusement Park
and
Dance Pavilion and Picnic Ground
and Other Stuff

If you live anywhere within six thousand miles of Big Lake, Minnesota you have probably already been to the Big Lake Amusement Park and Pavilion and Picnic Grounds several or many times.

If you have not, ask your folks to take you there.

If you live in Big Lake, Minnesota you will go there even if your parents do not.

But, bring your money that you have earned during the week working for the rich people who live around the lake in summer time only.

This would be done by "jobs."

Such as mowing and raking lawns or whatever they want you to do if you are strong and smile a lot and keep your eyes down as you will be shy.

To get to the Big Lake Amusement Park and all that you will want to take the path through Brown's Woods which will take you right to it unless you turn left at the fork.

In that case you will be at Brown's Beach but you can still get to the Park from there.

Just turn right at the sandy road.

You should try to decide before you get there what you are going to do with your money as there are so many more ways to spend it on than you can ever do on 20 or 25 cents if you have that much by doing odd jobs.

Almost everything cost you five cents or a nickel as we call it in Big Lake.

Before you get five feet off the road into the Park you will want to look over the first stand which will sell you little rubber balls for prizes that the milk bottles are setting on.

You must stand on your side of the counter and toss the balls into the milk bottle setting on the prize you want.

If you cannot see over the counter don't worry because the man will put his box out there for you to stand on so you get a good angle on it.

Or if there is just a number on the milk bottle, which have only small holes on their tops because they are real milk bottles, you want to be sure he shows you the numbers on the prizes hanging on the walls because milk bottles will not set straight on dolls or even jack knives which I never won anyhow.

I never saw anybody win a jack knife and they are great big red ones, but go ahead if you want to try it.

If you don't want to toss little rubber balls inside milk bottles you can do what is called fishing in the next stand.

The same man will see you standing there. So to go fishing he will duck through the door between the milk bottle place and the fishing place and sell you a pole with a hook on a string.

Then he will start the metal fish going by you in the water trough and you must try to hook one.

They have jack knives there, too.

They are really nice jack knives.

The biggest prize you can get there is a dusty bear which has been there a long time because if you ever catch a fish with a number on it, it is for a tin whistle or a magic ring.

That is why that bear is so dusty.

Now go over to the Skillball alleys where you sling a wooden ball down a kind of hallway and at the end the ball goes up in the air and when it comes down you hope it gets in the right hole and if you do this enough times you get big numbers at the end and then you might get a prize for that.

Men who have been drinking beer in the Pavilion do this, but nobody else does.

Most of the time the man who sells you the balls for five for a nickel reads a book in a chair.

Then there is a shooting gallery where you shoot a real gun at ducks going by.

They are made of iron and go clank and fall down if you hit them.

There are rabbits, too and they just sit there and you have to shoot them on the nose and it will go ping.

They fall down, too.

There are bullseyes and smoking pipes.

He has a box you can stand on too and he is always yelling to win a dollie or a dime.

My Dad always takes the money which is better than dumb dolls by a lot of miles.

There is also the merry-go-round which plays beautiful music and if the man who runs it has been going over to the Pavilion regularly he will make the horses go very fast and you can play Cowboys and Indians if you and your friends are spaced right.

We make a lot of mouth noises while we are on the merry-go-round and you will want to yell or pretend you are jumping canyons and stuff.

It is well worth your nickel if he has made the trips to the Pavilion or leaves it running while he goes to the toilet.

His name is Pete and you will know him because his front teeth are gone. He has fallen off his own horses many times.

They also get a Ferris Wheel to come there on the 4th of July, but I have only seen it once.

It is best at night when you go up in the air and you can see all the lights and hear the music and noise on the ground in a circle as you go up and over and down and you do that several times before they stop your seat and make you get off or give them another ticket which is a dime.

As long as you are way up in the air you should look for your house if you live in town. My cousin Dick said he saw his house in Monti, but he is what they call a kidder, so I just don't know what to think.

If you have still got a nickel you should give it to Bill Blackhurst and he will let you go in and ride the bumper cars in which you bump each other all over the place on an iron floor.

The bumper cars have numbers on them and you must learn which ones go the fastest and can still back up.

Some of them are slow and don't back up and then Mr. Blackhurst has to shut off the electric valve and push the cars from out of the corners and then turn it on again.

This is a very good bargain and will get you in shape from snapping your head all over the place so

hang on to the handle tight which if you push forward you go frontwards and pull back to go backwards.

You probably already know how to make them
turn.

After that you will have a good loose neck and can
watch the men try to ring the bell by smacking a piece
of tire on a board with a great big mallet and the thing
goes up a tall thing with numbers and writing on it.

It is not really a mallet but what else would you
call it?

If they ring the bell everybody claps and cheers
and he'll give the man another nickel for three more
smacks or a cigar.

And ring tosses, but that is mostly for ladies and
girls so don't bother.

But, all of the amusements in the Amusement Park
the best is the Penny Arcade which is only a penny
each for anything in it so you can go in there with
only one penny and watch after you're through with
your own money.

Mr. Klien owns it and he is from Minneapolis and
he will sell you pennies out of his belt or any other
place he has them such as his penny booth.

Mr. Klien can fix anything and always does so
very quickly after somebody busts his machines.

There are many of these machines where you put
your penny such as the one to see how much electricity you can take before you keel over or yell or let go.

This one is very economical because you can make
a circle with your friends or even your enemies so

hold hands except as soon as somebody yells and jumps and lets go of your hand and then your penny is shot.

Whenever that happens everyone points and yells that they didn't do it. I have learned that the one who yells first did it.

I have often been the one to do this and have been called bad names by my friends and enemies, but so has everybody else.

If it was your penny you can yell back at them.

There is a weighing machine that tells your fortune at the same time and another one that gives you your fortune in a glass tube which you unroll and if you can't read yet get somebody who can.

Even if the glass tubes are short they are very nice to take home and suck up nectar or if you're lucky there will be some lemonade left over from supper but don't get caught sucking lemonade or nectar with your glass tube or your mother will have fits.

And, don't run with your tube in your mouth or you'll never hear the end of it.

Then there is the squeezing machine, which we are too young for because you can't squeeze hard enough to make the needle move and make any impression on your friends.

It is not worth the penny for that reason and if you did and people heard you grunting and jumping up and down trying to get the needle to move they would call you sissy and weakling or worse.

Mr. Klien has also got a glass machine that is full of prizes like compasses and rings and a watch which you can try to pick up with the derrick you move around with a handle and a wheel on the outside.

And a red jack knife.

What you do is jiggle the derrick around by the handle 'til you get it over the prize you want and then your friend who is on the other side of the glass machine and you have got it all lined up you drop the grabber down and hope you got what you were after and then carefully do your maneuvering so the grabber is right over the prizes chute and let it down and then drop it.

If you didn't get what you wanted yell at your friend and blame him because it will do you no good whatsoever to blame Mr. Klien.

There are also jaw-breaker and bubble gum balls machines or peanuts.

All of these are delicious even if they are not fresh and they never are but who cares?

If you are brave and don't care what people think, there are the movie machines which have movies of bathing beauties in some of them and you put your eyes into this looking device and turn the crank at any speed you want and you had better start with the boxing show or the comedy show at another machine to throw anybody who is watching you off the trail and then work your way to the bathing beauties machine.

It is very educational.

You had better do this while you are alone because if you are with older boys they will push you away so they can peek, too. That is why they are called peek shows. Jimmy Johnson calls them peep shows, but his brother Darwin says Jimmy is always wrong.

After you are through in the Penny Arcade you will want to go to the picnic grounds because you will need some nourishment.

If your family isn't there one of your friend's families will be with sandwiches and potato salad, baked beans, cole slaw, fried chicken and cake.

There will also be lemonade and nectar, which is often purple.

They will say that there is always room for one more and you would be mean if you didn't help them enjoy their picnic.

And eleven kinds of pie.

Or they will have their ice cream freezer and a block of Bill Glenn's or Mr. Glase's ice which you can help make into chips.

Ice cream is made of cream and eggs and what you call custard which they will have already mixed up at home and brought with them in a gallon jar or pail.

It is a good idea when selecting a picnic to attend to look for people who have a block of ice inside a couple of gunny sacks if you haven't already spotted their freezer under the picnic table.

They will always let you turn the crank for nine or ten hours and scrape the paddle at least.

This can be done at home, too, but the day you are visiting picnics will probably not be the day you will help make ice cream at home because your brothers will be at the Park too and there is no one to turn the crank there except your father and mother and they know that's your job.

Turning the crank.

But they will anyhow most of the time if they like you a lot and think you are a good worker.

By the time you have had some of their picnic with them the father of your friend will yawn and stretch and tell your friend's mother tht is was pretty good and to take her time cleaning up because it is time for the baseball game with Orrock or Becker or Clear Lake or Elk River or Nowthen or Santiago or Monticello.

Baseball is free if you don't have any money to put in the hat.

You can sit in the stands or in the shade of the trees which is probably better.

Many people do both for a while if there are no wood ticks.

It will be very hot in the stands and the board seat will burn you at first but you will get used to it or head for the shade.

Try both.

And hope it is Orrock that the Big Lake team is playing because they are your worst enemies unless it is Monticello who are your most deadly enemies in the world by a long ways.

Monticello has uniforms and red caps.

Orrock has also got some uniforms for some of the players and some different ones, too, mostly gray with blue or yellow caps.

Orrock is best because they are all Hansons from

the same family except one or two and the catcher who is Les Durr, America's best catcher and wrestler and boxer and weighs a lot without fat.

Big Lake fans always say that Les Durr is really from Becker and shouldn't be allowed to play for Orrock and that he is muscle-bound.

They are also scared of him when he is batting and spits on his hands.

Les Durr will hit about three thousand home runs every year.

The Hansons are all pitchers and can pitch with either left or right hands, most of them, I think.

You cannot tell one Hanson from another.

Big Lake has Lyle Smith at first base and he also has to tell all the others what to do because he is CAPTAIN.

Kendrick Knowlton is in charge of the game and he has the baseballs, but if Les Durr socks one over the road and into the woods there is always someone who will chase the ball and never come back so there could be some delay.

There is always a lot of excitement and what is known as errors.

An error in baseball is when a player doesn't catch a ball or throws it over somebody else's head, for the most part.

You will get the hang of it if you sit anywhere near the Kolbinger boys who were all world famous players in Sherburne County at one time and are also known as legends.

Jack Kolbinger was a pitcher at times, but was a

legend about throwing out players from left field who tried to get to home plate from third base after Jack Kolbinger had caught a fly ball and winged it home with only once bounce exactly three feet in front of the catcher every time.

People who brought beer from the Pavilion are not allowed to thrown their bottles on the field during the game, even against Monticello.

After about four hours you could go home because Big Lake would be either 20 runs ahead or 20 runs behind.

By this time the band would be at the Pavilion and you should always see them before you go home. Ask them if you could help them carry their stuff up to the bandstand.

Some will and some won't, but stick around because the piano player will play a couple of hundred tunes while he fixes the piano for the night and the pretty lady will hum and sing into the microphone while complaining about it.

When you get home for supper be sure to tell you mother and father about everything until they tell you that's nice.

Thank them for letting you go to the Amusement Park.

You probably won't be too hungry for supper, but that's okay.

Nickel or No Nickel

BATH HOUSE, BIG LAKE, MINN.

It is too bad if you were not in Big Lake during the depression for our summertime.

If you were a kid during the depression in Big Lake you would have a lota stuff to do whenever you had time and liked to walk or run.

Whatever you liked to do you could as long as it was free and you didn't cut across people's lawns to do it. If you cut across people's lawns they would advise you thoroughly on where the road was so use it.

There are two lakes in Big Lake. One is called Big Lake.

The other one is called Mitchell Lake.

They are both very cold, especially when it is 96 degrees or even 93 degrees.

There is not any inlet and there no outlet either. They are spring fed and if this is so, where does the water go? I don't know and never have.

However, some people say that Big Lake has got no bottom. I don't know what that means either, but maybe that is where the water goes.

When it is that hot and there is no shade to be found about noon and the sun must be directly over-head but you don't dare look up because you will definitely go blind it is time to make some serious plans about what to do with your overheated body, namely head for Big Lake and go swimming if you have a nickel or will walk an extra mile to the free beach at Mitchell Lake which the Jaspersons let kids use.

Now, you know you are barefoot because shoes in

summer has never been practical or even useful and I don't know why you put up with it now.

If you have got shoes on you aren't ready for summer or you are a rich kid from Minneapolis and don't know any better.

Your feet have got to be well-seasoned and you would have done this in May and June after school and weekends and during recess so you are ready.

You really must be ready because the dust is terribly hot on the road you have to shuffle on and then when you get to the road with tar on it turn right but stick to the gravel on the shoulder or you will never get that stuff off your feet and ankles so you can go in the house again.

You should have also practiced gravel dancing too, because you are going to need it, as you can well imagine.

Don't go too soon after dinner or lunch as it is called in the cities, or you will get the cramps and sink.

The walking, running, shuffling, jumping and gravel dancing would usually take care of the time between dinner and swimming.

To get to the free beach on Mitchell Lake where it costs no money you want to go down Main Street just like you were headed for Brown's Beach on Big Lake where it costs a nickel. Nobody will know the difference whether you have got a nickel or not. If you go down the road past the side of the school house people will know you are going for the free beach where Jaspersons let you swim and use their raft.

Now when you get to walking down Main Street you would want to do that fairly fast so you don't spend the nickel if you have one. Garvers has a whole window full of penny candy and black licorice and baseball gum cards, and stick candy and jelly beans and candy corns and caramel and Smith Brothers cough drops by the box.

If you want to look, do it on your way home.

If you had two nickels the sky was the limit. You could go to Brown's Beach and stuff yourself silly by supper time on Bub Garver's candy.

You must also be careful on your way to go swimming so you don't get too close to the bum's bench where the old geezers spit on the sidewalk or say nasty things at you and laugh or take a swing at your dog to get him to yelp and run.

Look carefully as you cross the Jefferson Highway and it is okay to cut across past Brown's Hotel for the path through Brown's Woods. There is always a bees nest under the stile so don't poke around there too long or you'll wish you hadn't.

As soon as you're on the path you'll be glad because it is cool and sweet smelling from the honeysuckles and wild roses and cornflowers and the bluebells and foxgloves and even gooseberry bushes which are yellow.

You and your friend will want to scout the woods, too and look for money. I never found any but I heard somebody once did.

When you get to the fork in the path it will depend on your financial situation and your intentions. Take the branch to the left if you have the nickel and intend to give it to Bud Brown to go swimming at Brown's Beach.

If you don't have a nickel, take the branch to the right.

Now, you have got one more decision to make after you are on the branch to the right.

When you get to the Big Lake Pavilion you could go down the bank and swim free there if you have your swimming suit on under your pants.

But, you are barefoot and everyone pitched their beer bottles and whiskey bottles down the bank and they got busted. If they didn't get busted somebody else pegged rocks at them 'til they did get busted.

Sometimes you don't see them.

The swimming place is right next to Brown's Beach with a barber wire fence running out into the lake to keep you out. There are sixty seven million rocks there and there isn't a single one on Brown's Beach.

I don't know why the water doesn't seem as nice on your side of the fence either. Maybe from the beer and whiskey.

It has got water bugs, too and no place to lay down.

And, you are going to be seen by people who paid to go swimming.

It is always worthwhile to go out past the fence and swim under water over to Brown's, but you will get caught and Bud will tell your Dad. He is what they call merciless in the movies.

Maybe it is best to keep on going to Jasperson's where nobody looks like they paid and they didn't.

The knot holes are never in the right places in the girl's dressing room at the free beach.

Jaspersons will sell you candy and pop or near beer if you have any money but never charge a red cent for anything else.

Whatever you do on the dock is okay. You can run and jump or do the cannon ball or belly floppers or try to dive.

Diving will frequently grind the skin off your face and chest so dive shallow if you know how and keep your hands out in front of you to help avoid such inconveniences and subsequent consequences.

And check for rocks. And splinters on the dock.

The side stroke is a slick addition to the dog-paddle which is what that stroke really is when there are eighty-seven billion kids swimming at the same raft which is keep afloat more or less by the four barrels which seldom sink.

And don't try to haul up the anchor and do a little separate raft sailing or Mr.Jasperson will bellow at you and kick you in the rump without reservation or respect to your minority.

You will easily be able to tell when it is around five o'clock. Most of your friends and some total strangers will be blue. Some of this is from opening your eyes under water a lot while diving for rocks which you will have been throwing in the water to retrieve in a very interesting game which can be played in groups or alone which is very nice because if you haven't been finding the rocks your friends have been tossing for you to retrieve you will soon be playing alone anyhow.

Also there will be a good deal of standing in the warmer shallow water and your shoulders will turn inward across the chest for some reason. Lips are especially good time tellers and five-thirty lips is a condition which is unmistakable but not fatal other than at home where you will surely get remarked about for being late for supper.

The run home of five-thirty lips is more like a dash whereas you will have lots of time if you have left before you see all the blueness around you.

By the time you have swung your bathing suit three or four hundred thousand times you will be back in town and you will be dry and begin to notice how hot it still is and maybe how you are more than somewhat dusty all over again.

Now, if you have been able to take your nickel to Brown's Beach you are in a different situation altogether.

As you walk down the concrete steps into the bath house you will smell the minnow tanks and the toilets and you will see Bud Brown or his sister Lorene and if it is sister you will be nervous and in love with her. And the girls your age that have nickels will hang around the tanned Bud Brown and ask dumb questions like is the water cold today.

If you don't have a nickel you should have been up early to dig two dozen angle worms which is legal tender in Big Lake, Minnesota and therefore makes you eligible for entry into the bath house locker room and shower which is hard to understand and I don't know why anyone would take a shower before you jump in the lake.

Anyhow you will get a key on a rubber inner tube band and you put your worldly goods and chatels into the right locker and put the inner tube band over your head with the locker key.

The kids from the cities who always have nickels and are tan all over instead of garden-working tan will be the same ones who double the rubber inner tube bands and put them around their ankles like that is such a hot deal. And they will have cigarettes of their own which they hooked from their Dads and blow out their noses or make smoke rings with.

They have towels.

It is more better to snipe butts along the road.

But you cannot tear out on the dock where you walk at Brown's and observe the ropes for non-swimmers, whoever they are I don't know.

The slide you can go down regularly unless Bud is inside selling worms and then you go down in a header. And you can also try to climb up the waterwheel and from the wrong side or try to sit inside it or you can get underneath and hold it so nobody can get it turning.

You can also sidestroke out to the big raft where there are ten barrels holding it up and the diving board.

But look mostly at the diving tower as it is twelve feet high.

You can build up your ability by climbing up the wrong side of that and jumping off from say four feet at first and when you get some girls around you will be able to do higher.

It is much further looking down than up.

You must dive as jumping even without holding your nose is not at all notable and girls will just turn the other way on any guy who will hold his nose while jumping off the twelve foot tower.

You can also swim under the raft and fill your mouth with lake water and squirt it at the girls through the cracks in the raft who are laying there taking up room to get a tan when there is a whole beach by the bath house to do that and it is better for you anyhow.

Someimes you can get a girl to come under the raft with you and then, Oh Boy!

Whether or not you have a nickel the water will still pucker you up and turn you blue in four hours and thirty seven minutes so you have to go home.

Oh yes. If you gave Bud a nickel in the afternoon and you have to go swimming again after supper he will let you in free, but no locker.

What To Do With Rivers

When you have time in Big Lake, Minnesota there are a lot of good things you should do to enjoy rivers.

First of all, learn how to say Mississippi because that is the most fun to say.

And there are lots of ways to say it which you will quickly want to do.

It is also easier to spell than St. Francis.

But not easier than Elk River.

Which are the three rivers around you which you will want to do things with or to.

The two best things you can do to rivers is go fishing or swimming in them

This is how to do it.

But do not expect to go swimming in the Mississippi River because it is what is called swift and has eddies and drop-offs and undertows.

An eddy is where the water runs around a corner and goes backwards.

A drop-off is a hole in the bottom of the river.

Undertow is the worst because it will carry you out to sea under water.

Many older people get undertow and die.

So stay away from the Mississippi River when you are going swimming because you can never be too careful about drowning or being carried out to sea.

And if you are going fishing in the Mississippi River you will have to borrow a rod and reel from your older brother or your father.

With stout black line and a leader and some spoons.

A spoon is a Daredevil, but some have different

names.

In order to go fishing with a rod and reel you must take lessons from somebody who knows how to do casting and has time to teach you.

Your father is best.

He will take the Daredevil off the leader and put on a lead sinker and take you out in the backyard where there is plenty of room and no trees like the driveway.

Watch him and listen and when he is through talking he will give you his rod and reel

He will tell you how to keep your thumb on the reel and not press it down until the sinker is as far out as it wants to go.

Or you think the sinker has gone far enough.

You hold the rod in your right hand by the handle and bring it back over your shoulder and then let her fly, he says.

Sometimes putting your thumb down late will make a backlash almost every time.

Then you must unsnarl the line for the next two hours which will take up all your fishing time and make your Daredevil sink to the bottom and get caught in the rocks or weeds or trees which are being swept out to sea.

The trees are already dead but you can also catch your Daredevil in the trees that are alive on the banks or mostly behind you.

So practice everyday until you get it right.

In your backyard.

Shoot for targets.

Your Dad will be pleased when you learn how to do this really good.

Because it is his rod and reel and black line and leader and Daredevil.

Learn how to tie leaders onto the black line in case you break it on a great big fish called a Northern.

Sharks in the Mississippi will snap your line everytime and bite through the leaders and that is just as well if you have ever seen a picture of a shark.

Take along an extra leader and an extra Daredevil in your pencil box from school.

That way you won't get the Daredevil caught in your overalls or worse.

You won't be in school when you go fishing in the Mississippi.

If your mother won't let you use your pencil box go to August Peterson's Drug Store and ask Mr. Peterson if you could please have a cigar box which is empty.

He will always give you one and a fat rubber band to keep it shut.

You can now walk to the Mississippi River which is two miles away if you know the shortcut through Mr. Angstman's pasture.

Mr. Angstman will always let you go to the river through the pasture as long as you don't open any gates, but ask Mr. Angstman anyhow.

Then go past the barnyard down to the creek and wade in it 'til you get to the river.

The creek runs downstream.

When you get there set you cigar box on a rock

where you can find it.

Put your paper sack with sandwiches in it in the same place, but in the shade so they won't rot.

Put your nectar in the creek and make sure to put a rock on it so with won't float away and out to sea.

The nectar will get good and cold.

Now argue with your friend about who will get to fish from the big boulder and who will wade out on the sandbar where the creek goes into the river.

Always cast your Daredevil into the river either upsteam or downstream from where you are standing.

There are never any fish right in front of you.

If you are the one fishing from the sandbar and you get real hot, sit down in the water but make sure you don't have something important in your overall pockets like matches for survival.

You will always dry off by the time you get home.

If it seems like noon, eat your sandwich.

It will be ten o'clock.

What you will be fishing for will be Northern Pike and when they eat your Daredevil you will know it.

If you don't catch one right away yell at your friend that you had a strike.

He will do the same.

A strike is when a Northern Pike tries to swallow your Daredevil but misses.

Any bump or quick tug is a strike.

Or it will be the bottom of the river.

Yell about getting a strike anyhow.

You will catch a Northern Pike and whoever

catches the first one has to change fishing spots with the one who hasn't caught one yet.

It is only fair.

When you have eaten everything and finished all the nectar you may as well go home.

If you have caught any fish you will want to go right on past your place and walk through town with it or them.

Take the fish to Mr. Kobinger or Mr. Draves who will weigh it or them on their meat scales which are very accurate since they are both butchers.

If you are not tired from carrying them home from the Mississippi River they are not big enough to get weighed.

Put your fish in a tub of cold water when you get home.

Sometimes they come back to life but most of the time they don't make it.

Your father will teach you how to clean a Northern Pike.

I will not describe that.

Now you know how to go fishing in the Mississippi.

But you don't know about the Elk River or the Saint Francis.

If you are smart you will head for the place in the meadow where the Saint Francis runs into the Elk River.

It is about three miles away and you can usually hitch a ride as far as the town dump which is where you can get down to the Elk River without permission.

Go across the bridge and walk upstream through the woods by the cowpath.

Always be careful walking in cowpaths barefoot.

When you get to the edge of the woods crawl through the barber wire fence and help your friend through by taking his rod and reel from over the top.

If you don't go over the top and try to get it through the fence the Daredevil will get loose and wrap itself around the barber wire.

My Dad says you can catch anything with a Daredevil but a barber wire fence is bad trouble.

Now you are in the meadow and the grass is very tall and smells good in the hot sun.

When it gets to July the man who owns the meadow will come with his horses Babe and Maggie and cut it with his mower by sitting in an iron seat and steering his horses so all the grass gets cut.

Then he will hope it doesn't rain and come back in a couple of days and Babe and Maggie will pull a rake which he sits on in another iron seat.

He pulls a lever and it makes a pile of hay.

He keeps this up until he is done and then comes back with Babe and Maggie and a great big wagon which he loads with the hay.

His son John, who is your friend in school, helps him then and sometimes Doris who is John's sister.

Their real name is Jennings and they are all good workers.

If you go to Jenning's house to see John, Mrs. Jennings will make fried egg sandwiches with mustard on them.

Which doesn't taste the same anywhere else because she fries her eggs with butter she makes herself in quart jars by having Doris shake it 'til it is done. Doris is large for her age and very good at shaking.

They have a spring house where they keep cream and milk and butter.all from two cows.

A spring house is a house made of rocks over a cold spring that comes out of their hill and the water runs down to the creek which goes into the Elk River.

After the Jennings have finished with enough loads of hay they make hay stacks all over the meadow.

You might want to rest before you go fishing or swimming and the best place to do this is on top of a hay stack.

Take your sandwiches because once you get on top of a hay stack you will never want to get off.

Leave your rod and reel and cigar box at the bottom of the haystack.

A hay stack is big enough for one or two or even three but your friends will want their own hay stack to rest on.

You can see a long ways but mostly you will want to lay on your back and watch the clouds go by while you chew on the hay which is sweet at the bottom end.

When you are rested or wake up you just slide down the haystack on the seat of your overalls and head toward where the St. Francis River and the Elk River come together.

There are no sharks in either one but plenty of Northern Pike or smaller fish that are called Rock Bass. They are called Rock Bass because they taste like the bottom of a rock if you ever tried to eat one.

You can cast from the bank and catch them in the deep hole which is downstream.

Or you can get into the river at the sand bar and wade the shallow places and cast either upstream or downstream because there are no fish right in front of you there either.

The fish also go into their hiding places when it gets hot in the afternoon so there is no excuse for not going swimming at that time.

Put your rod and reel and your clothes in a place where they can be found.

Jump in the river.

River swimming is very different than lake swimming because you don't need a swimming suit which is usually needed at a lake unless you go after dark and many people do even with the mosquitoes by the bushel. It is called skinny dipping because that you are really dipping your skin, I think.

It is also different because you can swim upstream, which is hard, or downstream which is easy and a lot of fun because you can never tell whats under water in a river going between your legs.

Look up the river and you will see a very large tree bent over the river with branches going almost all the way across.

Head for that.

Climb out of the water and out on a limb you like.

Now jump or dive but if you dive you want to dive shallow and then only if you have made sure there is nothing floating by.

You must also holler when you jump or dive off.

What you holler is look at me.

If you are on one of the top limbs about one hundred fifty feet up your friends will look at you like you hollered at them to do.

Ear aches often keep your friends off those high branches when it is their turn, but not the lower ones.

When the shadows of the trees are all the way across the river it is time to get your overalls back on and head for home.

And it is a much longer walk back than when you came fishing and swimming.

Put your father's rod and reel in the corner of the back room by the ice box where it belongs.

Hug your mother around the middle and say what's for supper?

You will be hungry and she will grin at you and put down her spoon and hug you back.

How to Make Money So You Can Go to August Peterson's Drug Store and Ice Cream Parlor in Big Lake, Minnesota and Enjoy Free Movies

To get money for August Peterson's Drug Store and Ice Cream Parlor you must first find a rich lake person who needs to have his jobs done and will pay you.

The best way to do that is to walk around Big Lake as often as necessary and ask people who are out in their yard and look tired.

Say do you have any work for me?

If they are tired enough they will say yes and tell you what to do.

If no one is in the yard looking tired go up to their door and knock and then ask whoever comes to the door the same thing.

If a kid comes to the door ask if his mother is home.

That is because even if they have a kid it is anywhere from seldom to never that they would be caught working in the yard on doing a good job.

Wear patched clothes.

You will be surprised at the results which could run as high as 25 cents to more and up to a dollar a week.

August Peterson's ice cream cones are a nickel but they are big ones and well worth the money.

You eat ice cream with your tongue going all around it from the side and lick towards the top, then the top and start over.

Do not eat ice cream in front of your dog.

Another way to get good money is to take your gunny sack and hike toward Salida along the Northern Pacific tracks and spot iron and babbit from trains.

Walk slow or you could miss the babbit which is in clumps with a kind of gauze around it.

That gauze cloth is called packing and as you know it is for babbit in the train wheels.

I don't know what it does besides melt and run out with the gauze rags.

Babbit is very valuable and iron isn't as good for money but it will sell to the junk man.

Pick up bolts and nuts and anything that is loose but do not take the plates of iron that are stacked for repairs or the Section Boss Herman Kushman will know and come to your house and want to talk to your Dad.

On the way back toward town do the same thing on the Great Northern tracks.

Do not go to the other side of the depot and water tank as that is the tracks for the Nybergs who do the same thing and are bigger.

Take your stuff to Bill Glenn the ice man who does junk, too.

He will weigh it and give you money.

And, collect bottles which have which have been used for all kinds of pop. They are along the road and in alleys behind houses and stores.

The best place is by the Big Lake Amusement Park where rich people leave them or put them in the trash.

Dig all the way through it because they are probably on the bottom.

When you have got five that is the same as a nickel.

Which is what an ice cream cone costs. If you have more bottles Mr. Peterson will take all you have got because a root beer float is ten cents and so is that fizzly stuff they call "fosfate."

Mr. Peterson keeps his root beer mugs in the ice cream freezer and they are icy and you can burn your lips.

For some reason he will give your dog ice cream but you have to buy your own. My Dad says maybe Mr. Peterson just likes dogs and they have no nickels.

Sometimes during the summer August Peterson will bring Slim Jim, the Vagbond Kid and Don to his ice cream parlor part of his drug store and they will sing and play their guitars and accordians.

It is free and then they will have a talent show in which girls try singing and tap dancing and dress all up.

Slim Jim will crown the winner and also sell lots of snoose to adults.

Or Beechnut chewing tobacco.

Or Cleopatra cigars.

Or Wings cigarettes which are awful compared to Sensations.

Take my word for it.

On Saturday nights in Big Lake all three stores and August Peterson will stay open until nine pm at which time everybody takes their boxes or camp stools or whatever to the free movies in the vacant lot between Roman's Cottage Inn and Meeker's Cafe,

There is no rule against rich kids from the lake coming and plunking themselves right up front and making rude noises but there should be.

The movies are free and they are cowboy movies with Tom Mix in all of them.

The man who brings the movies has got only one machine so there is plenty of time to swat the six hundred million mosquitoes and June bugs who come there to bite you.

Some people say a mosquito can smell you for two miles.

Some of them come from Elk River and the people from Elk River say because you can smell the free movie crowd for ten miles

I don't think that is true, but you can't tell for sure.

After the movies you will want to do a lot of galloping and swatting your rump for more speed and yelling about all the Indians.

Don't cut through the woods on the way home.

And tell your folks about the movie so they can enjoy it too.

They will smile and shake their heads

Puppies and Kittens;
Cats and Dogs

Big Lake, Minnesota has a population of 417 people.

And thirteen million dogs and cats.

There are three kinds of dogs and cats:

Inside

Outside

Inside/outside

Your family will argue about what kind of dog or cat you have.

There is no argument if you are lucky and your cat has kittens or your dog has puppies.

These are all inside animals.

That condition only lasts a little while.

It is subject to change without notice.

Both kittens and puppies want to be classified as outside until they are old enough to qualify and then they change their mind.

Other things which change their minds and their position are:

Poor attitudes

Cold or hot

Scratching on furniture

Chewing up the house

Peeing

There is a lot to be said in favor of kittens on the last one since their mothers won't let them do that.

Don't ask where they go because they don't.

The kittens will scratch up a lot of mohair and other such furniture as they must practice clawing and climbing every day to achieve acceptable levels of proficiency, whatever that means.

A puppy will look sorry while he is busy eating your house and everything in it.

Their favorite food is old shoes.

They can eat many shoes in pairs or as singles every day for six months.

At that time they will graduate to larger household items such as garbage cans, tents, garage doors and neighborhood children, but they soon get tired of all of them and spend their time at sleeping

They will also slobber you up real good in the meantime.

Or any other time they happen to remember to do it.

You will learn to take warning from the mushy look they get in their eyes just before they come after you.

Hunting dogs as they are called never get over it and will jump on you to see if they can knock you down so they can get after you and do some very fast slobbering before you can cover up.

That is what they are hunting for.

When kittens are born they do so with their eyes closed and stay that way for two weeks.

After that they try to catch up on what they missed seeing and will tear all over to peer at everything in sight.

Kittens are very cautious about their environment and are suspicious of such things as chairs, tables, refrigerators, stoves, couches and everything that moves.

When a sharp noise happens a kitten will jump six

feet straight up from a dead stop.

Some kittens do not require a sharp noise to make them jump and will do that just to stay in practice or to tease other kittens, dogs, puppies or your Aunt Ida who is quite calm, ordinarily.

Kittens are extremely fast but have poor brakes and will often forget to lower their heads while dashing under chairs and couches.

They will continue to bonk their heads.

That is the sound their head makes on wood or metal.

A puppy is nowhere near as fast and has no brakes at all.

They are very fond of bonking their heads and will often repeat their bonking experience many times per minute.

However, a good dog is very important to growing up and reaching old age successfully.

Dogs are very proud of their achievements and their family background.

A dog's background is called by different names, but here are a few so you know what to expect:

Mutt

Mongrel

German Shepherd

Dogs such as Laborador, Collie or Shepherd are Mongrels.

Dogs which are called goofy names are all Mutts.

These would be rich people's dogs and would be poodles, terriers and that's it.

Kittens and cats are all from one family and they

are called alley cats which are farm cats who live in town.

Games with your animals, which are called pets, are hide and seek and fetch.

Hide and seek is especially fun with kittens as they are very serious about finding you and will perform leaps of great height when they do.

That is why there is a name Scaredy Cats.

A puppy tends to forget what he is looking for and is easily distracted by anything or nothing.

When you crawl out of your hiding place to see what he is doing he will greet you like he has misplaced you.

A puppy is hard to train to play fetch because of his short attention span.

They will often run to fetch whatever you threw but get lost on the way back.

Or cause irritation by refusing to give back your object for throwing.

Eventually they will catch on so don't worry about it.

A kitten or cat will only look at you and humm when you throw stuff for them to catch.

They are seldom interested, but would never think of holding that against you for being dumb enough to think they would chase something and bring it back.

A kitten which catches his first mouse will bring him back to the house and pataooee it out on your back step.

You should praise him as that is what he has in his

head to do.

If you praise him and take the mouse and toss it in the woods he will show up the next day with the same mouse and expect to be praised.

Keep it up and compliment him on his good work.

A sturdy mouse lasts about four days so that is not so bad for all the pleasure you have given your kitten.

How to Become a World Famous
Basketball Player
in Big Lake, Minnesota

When you are in third grade you will want to start working hard on your basketball game as this will make you world famous and maybe get to play for the Big Lake High School team and go to other towns on a bus.

There is only one bus that Dick Rosengren gets to drive because he owns it.

Then there is another one which is owned by LeRoy Larson who grins a lot and also owns a garage like Dick Rosengren.

People say you'd have to own a garage to keep them running.

But, before you can do anything you must have a place to practice.

If you are lucky your Dad will help you put up boards and any kind of hoop you can find, on the garage.

When it gets dark and if you have a yard light you can improve your night game by using that.

In the dark there are different rules than in the light.

You must choose up sides and learn to argue about the score at a moment's notice or any other time.

About six or ten to a side is good unless you have too few, but you won't because not many kids will have a bankboard and hoop and a ball and a yard light.

It is best if you have a ball that bounces so you can try dribbling a lot.

If not, any ball will do except a baseball.

A kittenball is okay, but a rubber ball or tennis ball is best.

There is no way anyone is going to have a basketball but that is okay because you'll have time for that later in your life before you become a world famous basketball player.

Since you are playing on the driveway you will want some rules on throwing rocks or gravel at each other.

Learn to yell, "you traveled!" even if your enemy is just slipping on the gravel as this will make him mad and he will throw the ball at you.

Catch it and it will be your ball and a chance to score or make a fantastic pass.

After dark it is hard to tell your team from the enemy and the rules against tripping, pushing and hacking do not apply as much if at all.

You must practice your shooting skills every day even if there is no game.

You should play all the time all year round.

If it is a Saturday, Bud McPherson the janitor will let you go into the school gym with him and even teach you a lot of stuff.

Bud McPherson is already world famous and can shoot with either hand or both and pass the ball behind his back anywhere he wants.

And he is known as a great janitor too.

And baseball player on the town team.

His wife is a teller in the bank, but I don't have any idea what they let her do besides tell, whatever that is.

Now you are not supposed to be there, but Mr. Ewing, who is Thomas Ewing's father, is the head janitor and Thomas wants to play just as much as you do and his father will even get an old ball out and pump it up for you the day before and leave it where Bud can get it for you.

Now you are just fine as can be because you have got a real floor, real baskets and a real ball.

But no shoes, of course.

Be sure you are wearing good socks but not wool or you will never be able to stop as good as with cotton.

Bud will have a whistle and tell you how to do everything like passing and shooting and dribbling, but be careful not to dribble much or he will kick your butt and tell you to pass, pass, pass.

He will also decide if you have skidded far enough to be traveling as you must know this or you will never make it.

You must always wave your arms a lot unless you have the ball.

If you have the ball always look for an open man.

That's what they call it if somebody goes to sleep as Bud says or has stopped chasing the open man to pull up his socks.

Bubby Hadden has got shoes.

His Dad owns the Citizens State Bank where Mrs. McPherson works as a teller.

Bud will teach you to shoot chest shots so you don't have to sling it from between your legs unless you forgot.

Most chest shots don't go very far.

Some Saturdays you might not be able to get into the gym and when it snows you can't practice outside very good because of mittens.

Go over to Keller's who have a barn and hoops in the hay loft and a big rubber ball.

You will have to clear a place down the middle so the hay isn't too deep.

Leo Keller is older, but no bigger, but it is their barn and he will make you practice lay-ups.

Keep your coat on until you are warmed up.

Then take it off.

And your mittens.

You may want to get rid of your sweater, too as you get hot from all the running.

It will be dusty.

Nobody knows where dust comes from, it just comes.

You can't get hurt in a hay mow unless the door is open on one end where they put the hay in.

Another rule is no pushing people out the big door because it is at least a hundred feet up.

And frozen ground below.

It is a good rule to follow especially if you get stuck guarding Leo Keller.

Glenn Keller can shoot chest shots and make them.

He is a husky player and bigger than Leo so try to be on his side.

By the time you are eleven years old you will be a wonderful player and ready for anything unless you have snuffed up too much hay dust or gravel dust or broken something you will need.

Go to all the high school games and act like you know what's going on.

Brown's Hotel
in Big Lake, Minnesota

If you have motored all the way from the cities on the Jefferson Highway to Big Lake, Minnesota, then you will want to go to Brown's Hotel for a nice rest and some good food as they have both.

You will see Brown's Hotel on the right hand side of the road just at the corner where if you turned to the left you would end up at the Saron Lutheran Church or Monticello. The first one is a nice thing to do but not the second. If you go across the bridge of the Mississippi River you will find yourself in Monticello and there could be trouble. Mr. Evers says the only good thing to come from Monticello was Harry Gahr and some others who moved to Big Lake because they couldn't stand it there any longer. Mr. Harry Gahr is a lumber baron and very rich and smokes ten cent cigars to prove it. Mr. Ever's son-in-law, Ray Warren, whose son is Ookie Warren, says that the people over there are snotty and stuck up and he is a Deputy Sheriff and he knows all about people as he has to deal with the public.

Anyhow, if you want to stay at Brown's Hotel you will want to go in the front door because if you go in the back door you will be in the kitchen and they will have to spend a lot of time explaining things about how to get to the proper place in the front of the hotel. You won't have to worry about getting a room there because they always have rooms to let whether you are going to be there for one night or many, which is how George and Edward Everett take their rooms and they have been there for a long time as permanent guests when they aren't living in the house in the back yard of Mickey McGuire whose sister is Midge because she is small. They also sometimes live wherever Mr. Larson lives and Helen, who is Mr. Larson's daughter and really pretty. She says that they are very nice men except she can't figure out why Mr. George Everett eats ice cream cones with his jackknife.

There are two things to remember if you stay at Brown's Hotel in Big Lake, Minnesota and they are both very important to your survival and well-being as they require you to pretty much know what the weather is like downstairs. If it is hot, leave your door open and if it is cold you should leave your door closed and open the register or you will freeze before you get to bed. You can always tell what the weather is like in Big Lake, Minnesota because there is a full-time weather man whose name is Ted Northrup who will make sure you are aware of the kind of day it is because he is always asking friends and strangers, "How's the weather?" I think he knows but he is not taking any chances.

There is plenty of free parking over to the left of Brown's Hotel, but do not park in the way of others, especially Vic Foote. Vic Foote is a permanent guest at the hotel, too, but that is mostly because he doesn't want to go out to his gas station at Bailey's Station because he thinks Bailey's Station should be named Foote's Station as he is the only one who lives there. Vic Foote has done a lot of traveling with some circus or other and he has got a yellow La Salle coupe with a rumble seat and he will let the older kids shoot crows on the fly while he tears around the gravel roads at sixty per if you catch him in the mood. If anybody is parking in the parking lot at Brown's Hotel where he thinks he would like to park his yellow La Salle even if he has got kids in the rumble seat he will swerve around making dust until somebody does something about it.

If it is hot out Brown's Hotel is the place to cool off as it is very shady around there and there is plenty of room on the front porch, which has rockers set up and somebody is always willing to sit and rock and talk about the good old days when the Red River Ox carts ran past the door and the Putman House east of town where the Nelson farm was was the only stop in Big Lake, Minnesota although that seems hard to believe because Big Lake, Minnesota has always had plenty of reasons for people to stop. If you cannot get a conversation going on the front porch of Brown's hotel then there is something wrong because most of the people spend a lot of time there developing their skills at talking on many difficult subjects.

Many of these same people are the ones who take their meals there in the big dining room even if they don't have a room there. Mae Brown is the lady who runs the hotel and she serves good food and meals from her kitchen where there is always something simmering or baking or cooking away. In the winter time when you come in the door it is enough to make your head swim because of all the good smells coming out. Mae Brown buys all her meat at Kolbinger and Draves Meat Market at 8:58 on Saturday nights so she is sure to get their full attention as they had intended to go home at 9:00. She is a kind lady who will buy your gooseberries even though you picked them in her woods at the back of the hotel and she makes them into what is called gooseberry pie.

They have every other kind of pie in the world there and their Chocolate Custard Rich Cream pie with meringue on top is so good people come all the way from Becker just to get some. If you like hot apple pie or hot mince pie with ice cream oozing all over the top you can get that because she will send her sister across the Jefferson Highway to August Peterson's Drug Store and Ice Cream Parlor to get you some fresh and then you will be able to say, "ooohhh" and "aaaahhh" and "I really shouldn't eat this after such a big meal" and other stuff like that. Also, for the main feature of the meal, Mae says nothing smells better than a roast in the oven and she means either beef or pork or chicken or ham. Nobody ever argues with Mae Brown because she is the boss, but there are plenty of people around Big Lake who think maybe her hot biscuits or butter top rye bread smells even better.

If you are a good eater you will get along very good with Mae and all the others who sit around and watch you while they and you are eating. They never say anything if you don't finish your plate like they do, but they will nod at one another and I think that means something. Especially if you order your breakfast of hot cakes and home-made sausage from Jack Kolbinger's secret German recipe from German Hill where he used to live and snuck it out when he married Bertha, his beautiful wife who can play the piano and organ and does at the Catholic Church in Becker and Elk River, depending on the priests. You must be industrious about your eating as they will give you fresh genuine maple syrup with Land-O-Lakes butter in a nice wooden tub and lots of home made toasted bread and some bacon which has been smoked for three weeks, I think. Mae's sister is a champion jelly and jam maker besides being quick across the Jefferson Highway and has got ribbons for both from the Sherburne County Fair.

My dad says they still know how to make coffee there, too, and he likes to put lots of cream in his and maybe have two or three cups after supper. You have to sit for quite a while after you finish in the dining room because you will find it hard to move fast or get up quickly. Sometimes Mae will have had one of her relatives busy on the ice cream freezer making fresh ice-cream out behind the kitchen and then no one will have to run across the Jefferson Highway to get some because it will be with lumps of fresh fruit in it, too. Mostly vanilla or chocolate or strawberry or peach depending on what she was able to get from Ozzie Warnecke in the other store or maybe even from Charlie Beech's General Store, but mostly he sells canned goods and crackers.

When you can get up enough energy to get up the stairs to bed everyone will say good night and then you can stretch out on the bed where you will sort of roll to the center and you can hear the birds outside the window getting ready for bed too and once in a while a motor car will go by and then in the morning after you have had the best sleep of your life the robins and blue birds will wake you up and the sun will be coming in the window and the lake breeze will be moving the lacy curtains around very cheerfully.

When you come down the stairs everything will be ready for you and you can start eating right where you left off and if it is during the school time the school teachers will be there to have their breakfast as they will tell you they need all the strength they can get because of all the Norwegians. They will nod to you and say good morning and you know you have come to the right place when you came to Brown's Hotel in Big Lake, Minnesota.

Good Eats

STREET SCENE, BIG LAKE, MINN

If you like to eat good food at reasonable prices you have many places to chose from in Big Lake Minnesota. There are three besides Brown's Hotel and I have already told you about how to eat there at Brown's Hotel so I won't tell you about that anymore even though it is very popular and has good smelling food all the time.

If you look next to the garage on Jefferson Highway across from Brown's Hotel you will see a great big sign that says Cafe. That is the name of Meekers and they have what are called specials. I don't know if they are special because of the way they are made or because they have special prices on them because they have made too much and have to sell it the same day that they say it is their special.

No matter the reason why, they are also special because of the way they taste. You can either eat at the counter or at one of the tables or even go into what is known there as a booth. They have some nice girls and ladies there who will take your order and go back to the kitchen and tell the cook what you want to eat and then bring it to you when the cook says it is ready. It doesn't matter where you sit because the tastes the same no matter where you eat it and that is that it is so good your eyes will water.

Some things they like to sell as special are roast beef with mashed potatoes and gravy. They also put some nice corn on your plate and maybe something else and a hot buttered roll. You can get coffee or tea to drink with it or if you aren't old enough to drink one of those you can have milk. If you don't want that because you can have that at home all the time and you have another nickel you can get a bottle of pop. They have many different kinds and colors and flavors of pop but most of the time you should choose grape because it will make your tongue and parts of you lips purple and it burps sweet.

If you want something else besides roast beef they also have chicken and roast pork and maybe something else but I don't know what it would be. The chicken is sometimes roasted like a turkey or it is fried.

Oh, yes, I remember now. They have something they call chicken fried steak which means they take a piece of tough meat and slice it like a steak and then put all kinds of flour and salt and pepper on it and then fry it in lots of lard. When it is done it is golden brown like chicken and it isn't tough any more. I think what they do is what my mother does when she has to fix tough meat and that is that they take it and they beat it with what is called a meat tenderizer and that is like a hatchet with lots of little blades on it and you pound the steak with that for a quite a while or until it looks almost like hamburger but is still hanging together. Then they put the flour and salt and pepper on it and fry it like I said.

If you want the chicken you can get either what they call a quarter or a half and that means the quarter is really 35¢ and the half really is 50¢. It is delicious and in the summer time they serve some corn on the cob with it and mashed potatoes and some chicken gravy and fluffy dinner rolls they make back in the kitchen every day.

If you choose the roast pork you will get the same things on your plate as you do when you order the other things but you won't care because it is all so good. If you like pie you have come to the right place because Mrs. Moores makes the pies and those are Apple, Dutch Apple, Peach, Pumpkin, Blueberry, Lemon Meringue, Banana Cream, Chocolate and Raisin. You can also get them to put on some ice cream so it is just like going to the Lutheran Church Ice Cream Social where they do that every year. Everyone says they really shouldn't have a piece of pie after such a big meal but everyone has one anyhow because nobody can walk out of Meekers without eating Mrs. Moore's pie, especially her Dutch apple. I don't have any idea why they call it Dutch Apple because the apples all come from the orchard in Howard Lake where Mrs. Meeker goes to pick them in her white dress.

Mr. Meeker always shows up when it is time to pay and he will say that is 35¢ or that is 50¢ or whatever it is. He is very careful with money and always gives the right change. If you don't have any money to eat at Meekers Cafe you should go around to the back door and you can smell the good smells free.

Next door to Meekers Cafe is a vacant lot or parking lot and next to that is another wonderful place to eat and it is called The Cottage Inn. Mr. and Mrs. Vernon Roman own it and they both work there but especially Mrs. Roman who is called Hazel and she runs the kitchen and cooks and builds her world famous strawberry pies. It is a secret recipe because people are always asking for the recipe and she says it's a secret.

They have a beautiful dining room which has a great big fireplace in it and there are table clothes on all the tables and they are red and white checked. The waitresses are all dressed alike in white blouses and black skirts and they call you sir and ma'am.

People say that you should always try their fried chicken because it is the best in all of Minnesota and it is, too, but they have steaks that aren't fried in lard and you get a baked potato with it and a salad as big as a house but my dad says you better bring your money if you want to eat there.

Lots of the rich people from the lake eat there and they say that the food is better there than in Minneapolis and maybe even St. Paul. It is also the place where they always have the Junior-Senior Class Dinner where everybody there gets to eat something called chicken croquettes which are lumps of chicken made into little pyramids and served with mashed potatoes and green peas and rolls. My sister says they are very tasty and I think that means they are okay.

I think they serve beer in The Cottage Inn, too, because I have been roller skating past there when some of the lake people come out in their dress up clothes and they are kind of wobbly and giggling and acting silly.

If you go a little bit further down the street past the Union Church you will find Putt's Cafe and that is

where you want to go to get your hamburgers because they are the best in the world or anywhere else.
Some people say that is because they never clean their
grill and that captures all the flavors over the years.
The minute you walk in the door you will smell the
fried onions and the hamburger and the buns are
toasted on the grill, too, so they come out all golden
brown and crispy and soft on the inside. They put on
lots of pickle relish and catsup and mustard and
whatever else Mona Snyder has on hand when she
makes them.

They have also got booths and a counter and they
have got nickels and dime slot machines and they
don't care how old you are if you want to play them
and waste your money. Putt's Cafe means that it is
owned by Bill and Gladys Putman and they have
built it up from when it was the Oriole Cafe and that
was just a shack that served hot dogs and hamburgers
only. Now they have all kinds of sandwiches and
lunches and they have a soup machine behind the
counter where you pick out a can of soup and they
will open it and heat it and serve it with crackers for
10 ¢.

They have got what they call French Fries, too,
which you eat with your fingers by dipping the ends
in catsup and then sucking them in.

They have also got about a hundred kinds of pop
and they also have beer of all kinds. All of their kids
work there sometimes and they are Gene who is a
great basketball player, Mertis, Beverly, Donna, Jerry
and Kate. Mrs. Moores also makes the pies there and

does a lot of the other cooking and the pies there are the same as they are at Meekers only the pieces are bigger but you can still get ice cream put on them if you can afford it.

If you are friends with Gene he will let you go down the basement with him where they have a pool table in the store room and you can play free. There you will have to learn how to handle your cue stick differently because of all the beer and pop cases that are stacked around it and sometimes on top so you have to take them down and find a place to put them but be sure to put them back on the table when Gene tells you that you are through playing, which can be pretty quick on some days when you seem to be going real good.

Next door to Putt's Cafe is another garage where you can watch Mr. Matt Reintjes do his welding only don't look right at what he's welding or you will go blind. He is very kind about telling you to get out of there when he's welding or pounding on something with his big hambers. His wife is the lady who owns the post office and they have two kids and a mean Shetland pony who will bite you hard wherever he can get hold of you and that isn't funny.

Next door to the garage is a mysterious place called the Purple Parasol and they have got real live concrete parasols painted purple out in front and inside when sneak up and peek into the windows you will see people you may or may not know drinking and dancing but I don't think they eat much so I don't know what kind of food they might serve there if any.

That is all I can think of for now about the food in Big Lake, Minnesota at this time.

Bag Swings

If you are ever going to put up your bag swing you will want to use a Bemis bag. That will be the heavy kind that looks a lot like canvas but isn't. I don't have any idea what it is, but it is very strong and will take a lot of punishment, which is what you're after when you put up your bag swing.

Gunny sacks aren't strong enough.

You will also want to have a good stout rope to hang the bag. But first you have to have stuff to put in the bag. It is important that the product you put in the bag is not lumpy or hard. Select the material carefully whether it be sand, which is a lot too heavy and could contain rocks, or sawdust which is excellent if it is just sawdust and not mixed with pieces of wood.

Sand is not that good.

Sawdust is a little too light to get a good free swing start, but it is probably the best.

Both will retain moisture and soak up your pants after a rain or any other time it gets wet. Two gunny sacks, one inside the other will work. They also dry faster than your Bemis bag.

They also rot faster. Bemis bags could last a full season if you don't weigh over 56 pounds.

What you have to do is to fill your bag about one third full with the material you have selected.

If it is heavy material you will want to put the material in at the site of your tree branch from which you will want to hang your swing.

If you think about that part you will see the wisdom.

Take your binder twine and wrap it around the top of the bag and tie it.

Better yet, take and fold the top of the bag double and then wrap the twine as tight as you can and then tie a bunch of knots in it.

And better yet, if you can just get a knot tied in the top of the bag you will want to do that. Then you won't have to look around for any twine in the barn .

If you know where the twine is you could still use it below the knot.

That's up to you.

When you have the bag swing ready to put up you toss one end of your stout rope over the tree branch not to exceed twelve feet in height.

Be sure of your rope. It won't do if is cotton clothes line. And it stretches a lot and could injure your rump.

The bottom of the swing should be roughly ten to fourteen inches off the ground when you are strad-dling the swing but not swinging.

You will notice that some adjustment must be made and that is why I didn't tell you to tie the rope to the limb yet.

You are now in the crucial test program where you have to determine how much the branch is going to give and how much the rope will stretch.

I think you can still get good hemp rope.

When you have got it just right you can toss the rope over the branch two more times.

Make sure your rope is about 25 feet long.

Now tie the loose end of the rope around the trunk

of the tree, but not too tight so it kills the tree or you can't get it undone if you want to adjust the arc or flight, as your father calls it..

Go and get an empty barrel. Roll it to where you will need it as a launching platform.

Stand the barrel up so you can stand on the bottom. Put some boards on it anyhow.

Go get a kitchen chair and climb up on the barrel bottom with the boards on it.

Tell your little brother to pass up the bag swing.

If you haven't got a little brother anybody else will do.

Actually, you will be far better off if you just go get Donna Larson, since she is your neighbor and is what we call a tom-boy because she can do everything better than the rest of the boys you could name.

The bottom of the bag swing when fully extended should not touch the top of your platform or you will bang your head on the way back.

Hold the swing and then let it go.

Climb down and check the bag for height again. Remember, it is your rump.

Now climb back up on the barrel and yell at your brother to hurry up and give you your bag swing. With a little practice and a few promises he will soon learn to swing it up to you and you will catch it with one hand like a trapeze artist in the circus.

If you have gotten Donna Larson to help you, you had better ask her nicely to pass the bag up to you on the barrel or she will tell you to get it yourself.

Be sure you have enough material in the bag so

you don't slip off. You should have made sure of that a long time ago.

Grab the rope and leap into the air and straddle the bag swing. Let out a holler if it works.

Better yet, let out a Tarzan yell but do not try to beat your chest or you will kill yourself.

After a couple of times put your head back and look up at the sunshine through the tree leaves. Look at the white clouds and sing "Beautiful Dreamer."

Be sure to let Donna swing on it whenever she wants to because she is a good friend and your best neighbor that you will walk to school with when she lets you.

Do this in the morning before school, but forget the yell or your mother will come running out of the kitchen and tell you you better get scooting which means you better get walking.

Do it after school and on Saturdays and if your folks say okay, on Sundays.

Be nice to your little brother. You could let him swing too, but not much.

How to Play Pool

If you want to learn how to play pool you should be able to at Eddie's Pool Hall in Big Lake, Minnesota.

You will want to start fairly soon.

A soon as you get a nickel.

But for a nickel you have to do what they call rack the balls yourself.

There is only one table, but that is okay because the older boys go to Monticello.

The men in Eddie's only play cards.

Such as cribbage, whist, euchre, pinocle and poker for match sticks or real money, depending.

They do not play bridge, whatever that is.

And, they say the table is tilted and is not live.

I don't know what that means yet.

And there are only three cues.

One is real short because of one wall and the stove being closer to the table.

Always flip your nickel heads or tails to see who gets to start which is called the brake or break or something like that and who gets the good cue. When you get older you can lag for first shot but not the cue stick. I could explain to you what a lag is, but it is too hard to do and would take way too long to tell.

The other one curves a quite a bit to the left.

Take the two cues and roll them on the table to see which one is straight.

Straightest, I mean.

If it is your best friend you will want to share the stick.

This will also show whoever is with you that you

know what you are doing.

Then give Eddie Walters your nickel and rack the balls.

Eddie will not take anybody's nickel who isn't big enough to see over the edge and put at least one hand flat on the table.

Get somebody to reach over the table and pull on the light so you can see.

Don't ask Bill Glenn to pull on the light or he will swat you one because he is the town hump back and can't reach it either. And remember, he is also your iceman.

You won't be able to do that unless you are way late in starting your pool life.

Now you have to argue whether you will play rotation or what is known in Big Lake as 8 ball.

Rotation is better because 8 ball means you can only shoot either stripes or the balls without stripes.

The way you decide who shoots what depends on whether a striped ball or a ball with no stripes falls into the pocket first on the break.

Sometimes nothing falls into a pocket and then the next shooter gets to take any color or stripe he wants.

So play rotation.

And learn the colors real good so you can try to get one or more in a row.

You can also play partners which makes it a lot cheaper.

When you finish a game you can yell rack and Eddie won't rack the balls but he keeps the rack under the bar so nobody gets to cheat on a free game.

Unless Eddie is helping his customers with the beer bottles or glasses and then you can leave the nickel and take the rack.

Sometimes Eddie will just be helping himself with his own beer so he doesn't care where the rack is.

Some people say he wouldn't care if the rack was in Becker.

Which is west of Big Lake about eight miles.

When a whole bunch of kids are there at Eddie's Pool Hall the oldest ones get to play first.

Andy Uram is an older kid but he will play with you because there is no such thing as a nickel in his whole family.

He is so skinny you wonder how he can even lift the cue.

But Andy Uram is very tough and he is what is called a dead eyes.

Eddie says if Andy did arithmatic in school like he can add scores he could graduate in two years.

Andy Uram is eleven years old.

His brother John is even older and in High School and is a world famous basketball star who can shoot one handed.

And from the corner on the run, over his head. It is called a hook shot and is the very latest thing to do in basketball but don't try it or you could kill your-self.

Andy Uram will be a world famous basketball player, too because he can already make chest shots from halfway to kingdom come, as the coach Mr. Sandboe says.

Whoever loses, pays for the next game and it is not a bad idea to make everybody put their money on the table to make sure there will be a next game.

That is where the saying pooling your money comes from.

Do not yell or holler or argue real loud or the card players will kill you or make Eddie throw you out the back door.

Listen closely if the older boys are playing and watch their moves.

They will be able to teach you how to shoot with a cigarette in your mouth when you are older but Eddie won't cork you one because you are too young to smoke.

To be old enough to smoke cigarettes, or cigarette butts you must be able to put both hands on the table without going to tip-toe and not be afraid of being told on.

That is not really fair because Andy Uram is so skinny he could make both hands flat when he was six and he doesn't care who tells on him.

Smoking didn't stunt Andy Uram's growth.

When learning your pool you must study your enemy and learn the names of all the card players because you can never tell when you might need an adult to help you escape if you get too lucky when you are playing against the older boys.

Also watch out behind you if you are on a hard shot because your enemy or somebody else will push your cue from behind and you'll miss.

Scratch means you put in the cue ball instead of a

real ball.

If you rip the table you will never get in the door again even if Eddie patches it with his adhesive tape.

You will feel so dumb you won't want to play for three days.

Eddie has the coldest pop in town.

Because everybody drinks beer and the pop stays in the cooler longer.

Eddie will also fry you a hamburger but they aren't much for a nickel when he gets through with them.

You better let Eddie's wife whose name is Mrs. Walters fry your hamburger.

Or Lucille if you are old enough.

She will fry your onions good, too.

Do not put your pop bottle on the pool table or it will wreck the table even worse when you spill it.

And you will.

You must learn how to shoot your cue ball both up hill, down hill and sideways because of the slope.

If the ball is chipped that you are going to shoot at you will just have to guess which way it will go because it will go clunka-clunka-clunka and all you can do is watch.

Do not blast the ball because they will jump off the table and besides getting lost in the beer cases or down cellar Eddie will cuss you and take your stick away.

And there goes your nickel.

Watch out for strange boys from Monticello who will hang around and probably are stealers.

They are older looking and are what is called shifty-eyed and wear tennis shoes in summer and say rude things at you.

Or Oxfords.

Which means they already have money and lots of it.

Do not even talk to them, of course, or they will slough you one, or push you.

If that happens do not yell for Eddie or he will insult you.

Make your games last as long as possible which isn't hard to do because you seldom make the shot you meant to and you are learning.

Making a ball you were not after is called slop if it is your ball color.

If it is not your ball it counts for your enemy anyhow no matter how much you complain.

If your cue ball doesn't hit anything ask to take the shot over because you slipped.

Nobody ever let anybody do that but try anyhow.

You have to keep at least one foot on the floor when you are shooting.

That is all there is to learning to shoot pool.

How to Play Tackle Football

When you get to Big Lake, Minnesota you should start to look for big, flat fields without too many rocks or sand burrs near your house.

A few are okay but not too many.

And it should be as level as possible with maybe some soft grass to land on in many places.

The field should not have trees if this is possible and very few ravines or puddles in it.

And flat.

A flat field of grass is not hard to find in Big Lake, Minnesota, but most of them are already in use by cows or older boys.

Generally, you will get what is left over or maybe not in use because girls have come by and taken all the older boys away and they are somewhere else doing some giggling.

You will want this kind of arrangement for a place to play some tackle football which is very good for you and will help establish your reputation in a hurry.

It is not good for your school clothes so the games have to be played after you have changed your clothes at home and done your chores.

Chores is what you do so as to earn board and room at your house.

Board and room is the same as eating and sleeping, but a lot of other stuff is involved in it and I don't know what isn't included in it except you better not act smart as long as there is a roof over your head and food on the table.

Chores are like splitting kindling and taking it into the back porch and taking the dry kindling from the back porch to the kindling box in the kitchen.

Chores also include duties which means doing what you are asked to do or feeding the dog which is not work at all but let your mother think it is work for you even if she enjoys it, too and often fries extra potatoes or pancakes just to give the dog plenty.

The dogs name will be Max and he will be a German Shepard and will fight your battles for you in the woods against wild animals and other such creatures.

Creatures of the forest are never seen except by big dogs.

Anyhow, if you ever get everything done you will still have about enough time for a good game of tackle football at a place you have chosen.

If it is in a pasture be sure to scout for cow flops because they will make you slip everytime.

If the cow flops are dry or almost anyhow just give them a heave and they will go sailing out of harms way.

If it is in a vacant lot look for dog piles but do not touch.

Kick them out of the way or don't pay any attention to them.

Unless you land in one.

There is no explaining that to your mother so do the best you can on your pants and shirt with some dry grass.

A football field for tackle football is about the same size as you have available.

It is always longer when your side has the ball because somebody will move the goal line markers which causes arguments and a few fist fights.

Which is no worse than getting tackled by Ray Nelson.

When the football shows up it is time to choose up sides.

It would be best if you get two or three of the Johnson boys on your side but there are only four of them and you can't have more than two on one side because of the rules.

Also hope for at least one McPherson and one Keller and any one of them will put you in good shape.

And Ray Nelson of course.

And Thomas Ewing who can hike the ball eight-two feet which is what you want to give you time to get down the field for a pass.

The Big Lake All American Tackle Football Team would be hard to beat anywhere as they would be tough with such stars as I have named plus a few more I haven't.

The captain will always tell you what to do so listen closely or he will slough you one if you didn't do it when you get back to the huddle.

The huddle is where the captain draws out the next play for everyone.

Before the game starts look for any new gopher holes.

You will either kick off or get the ball.

You get four downs to score or you lose the ball to the enemy.

After the kick off the captain will draw the play

and that is what you must do.

And no tripping.

If Lefty McPherson is in the game look out for passes because he could heave one to you or against you so fast it will make your head spin.

The idea is to run the ball as much as you can but look out for a pass as the runner will get rid of the ball in order not to get killed on the spot by Bud Johnson or Ray Nelson.

When all the Johnsons are playing and there are two on each side look out because they will often just go after each other and beat each other up because they are not allowed to do that at home in the house as much as they'd like to.

There are many secret plays in tackle football that looks like they are accidents but they are really that secret.

For instance a secret play is when Thomas Ewing hikes the ball eight-two feet when the runner is only forty feet back or so and then that secret play is that everyone must quit what he was doing and go out for a pass and hope you catch it if it comes your way and it could because the runner is doing the best he can to get the football and sling it so he will not be tackled for a big loss.

Another rule is about spitting at people after they have scored on you.

So spit to one side and yell at somebody else for not doing their job.

Saturday afternoon is the best time to play tackle football in the fall right up until the snow makes you quit.

Many times the older boys will let you play if they are short a few players so hang around and watch to learn their way of playing.

Most of them will take it easy on you if you have got the ball by accident and they plan to tackle you. Mostly they will just grab you by the head or your bibber straps and sort of hold on to you until your feet stop.

Which is way better than Ray Nelson will do.

If you are playing with the older boys and if you ever make a touchdown by some mistake look out as Darwin Johnson will pick you up and toss you sixteen feet in the air and then catch you and the ball even if he's not on your side as Darwin really likes tossing people up in the air and seeing touchdowns make him very happy.

At school you shouldn't be playing tackle football in your school clothes so it will be touch football for a few minutes and then turn into tackle without adequate notice if you have the ball.

That is also very hard to explain when the knees go out of your good school pants and your mother has to patch them.

Always remember what your sister's boyfriend, Ernie, says about playing tackle football. He says there are bound to be casualties and sometimes it is your pants and sometimes it is you. Do not expect to come out of it the way you went in because that is expecting too much or you aren't playing hard enough.

It will help you to remember to comb and brush your mother's hair when she is patching clothes after supper dishes.

She will like you to braid it too so she will always say yes when you ask if she wants her hair fixed up nice.

How to Feed Rabbits, Ducks, Pigs and Fences

Glosssary of terms: Banure is Big Lake politeese for manure.

A sigh is a long-handled, long-bladed cutting device used on tall grass or like vegetation by ambitious people It is often misconstrued to be a "scythe".

I. Feeding Rabbits:

When you have gotten up to about four or five years old in Big Lake, Minnesota you will be just about the right height for feeding rabbits in the barn you will be working in.

When you are older, about seven, you will be tall enough to feed the rabbits in the second story hutches.

By that time you will be sick and tired of feeding rabbits so try to forget about that age.

Feeding will still be okay, but cleaning the hutches won't be because the tray for rabbit banure will be at eye level and will distract from most of what you plan to do in later life.

There are many kinds of food to give rabbits that will make them have baby rabbits after a while.

Most rabbit food is free and very nutritious for all stages of the rabbit making business.

The first thing to do is to watch what your father does when he cleans the hutches and their water bowls which are very heavy and made out of what looks like rock.

He will rinse them out very good before he puts in fresh well water from the pump except in winter when the pump is frozen.

In that case he will bring water from the house to give them or pour red hot water down the pump to make it warm up and feel well enough to give you water.

This way is more work but your father won't mind as he is always determined to do things the hard way as your mother will say.

Before you do watering in summer or winter you should clean the hutches by scraping with a long handled hoe into the tray under the wire through which they should do their duty only sometimes they forget or miss and go on the ledge they stand on to eat their meals through chicken wire which is called a feeder.

So it goes like this when you are still doing the first storey of rabbit hutches.

When you go into the business of feeding rabbits it also means cleaning up so don't think when your father says you can help feed the rabbits he means only feeding because it is all one thing including lifting heavy rock bowls, cleaning out the hutches, cleaning out the feeder and putting in new feed.

Your father will also have names and how old they are and what type of rabbit he has got in his hutches because he could be called a rabbit breeder which means he is willing to keep on feeding rabbits and cleaning up after them until they are fed enough to cause baby rabbits.

I think.

Anyhow, he will cut long grass and fresh clover with his sigh and carry it into the barn and flop it down in the grass and clover area but not in the rabbit food bin which has a cover over it and has rabbit pellets which he also gives them in another rock bowl. He never mentions to your mother that he bought pellets and he thinks she thinks he buys nothing for his rabbits, but you would be a fool to believe she doesn't know and fumes about it.

So that makes two rock bowls in each hutch except where there the little rabbits are after they are given their own hutch to live in when they are old enough, say four weeks, but are too young to chew up the pellets so that's out.

Before that the baby rabbits are very nice to pick up by the back of their necks and hold in your arms and nuzzle.

Nuzzling rabbits is a wonderful thing to do if their hutches are clean but not so great otherwise.

Your father will lots of times bring cute little rabbits in the house for everyone to nuzzle a little.

If you have lots of brothers and sisters your father will bring in plenty of baby rabbits so everybody can get in their share of baby rabbit nuzzling done.

That goes for baby chicks, too, except a baby chick, when they are yellow, are full of you know what and they are not safe for more than seven seconds per nuzzle before they have an accident

Never hold a baby chick above your head.

I will tell you about ducks later, but not now so hang on and pay attention to your father with rabbits.

Some rabbits which have names and birthdays on their hutches will be called by that name by your father and are bucks or does.

These are Chinchillas, New Zealand Whites and

Rhode Island Reds.

Forget Rhode Island Reds.

Those are chickens.

After your father has brought in the grass and clover and alfalfa and pig weed he will clean the hutches like I said and pull out the tin trays which will be fairly well filled up with what they ate the day before.

It does not smell like clover or alfalfa so it must be pig weed.

What is in the trays will clean out your sinuses, whatever they are, but believe your father and what he says it will do.

Do not take any deep breathes until you get out-side anyhow.

Then he will open the hutches and do the rock bowls and push in the green food.

They also like to eat a lot of lettuce which will have gotten too big for you to eat from the garden but pick it fresh and cut the dirt end off.

They don't care much about radish tops but like crab grass and spinach uncooked,

A rabbit will quit thumping when you feed him and stay quiet and call them by name.

A rabbit doesn't like huming or singing to them like a cow does when they are milked but that is because you don't have to milk a rabbit because the baby rabbits do that for you.

You put all the stuff from the trays outside in a pile where it mostly rots.

You can tell which rabbits have been shipped to other rabbit people or ones you ate when their names disappear.

That is all there is to know about rabbits for a while.

Except rabbit mittens your mother gives you in winter which I'm not sure about.

2. Ducks:

Everything a duck does is funny.

A duck cannot help it, so do not tease them all the time or they will not be your friend and a good duck is nice to be around and appreciates your company when you are alone with one or a small group of them.

A small group is two or three but four does not work.

Get to know your ducks while they are little and study the way they go about doing things.

A good duck does not need a fence to remind him where he belongs.

Or her.

Ducks can have many names but do not try to get smart with them by using dog names as they resent names like Rover and Spot or cat names like Tiger or Kitty although that one is pretty and fun to say.

A duck does not resent much, but used names will make them sulk a lot and waddle off and shake their feathers.

To get a duck to follow you, you should start by getting friendly and learn how to waddle like they do.

It is best to do this while the ducks are little because older ducks think you are just being silly when you do your waddling and will walk off and talk among themselves.

If you have an old tub or part of a barrel cut lengthwise left over you should get your brother to help you get it close to where you can put water into it.

If you don't have this but you can get a great big huge tractor tire and have somebody lay it flat and put water in the bottom half.

A duck or some ducks will swim around inside the tire for many hours and do some very nice quacking while they are at it. It will make you giggle a lot.

Do not let your brother anywhere near with the B-B gun he got for Christmas because all older brothers can think about is shooting birds and animals and will tell you a duck in a tire is good target practice like a shooting gallery at the Big Lake Amusement Park only free.

B-B guns are dangerous weapons and can put your eyes out, just ask your mother if you don't believe me.

Ducks also like puddles a great deal and if they are too shallow to paddle in they will stand in the water and catch bugs and gargle with them for hours.

A duck will do a lot of talking to their friends and often come up to you with a question in his eye like why don't you do this or that.

About the only thing you can do is guess what they are talking about and once in a while scoop them up if they want you to and sit in the shade and scratch their necks about where the bottom of the green color is.

They will fall asleep because they trust you as their friend.

If a lady duck likes you she will give you an egg which is much larger than a chicken egg and is a meal in itself as your mother will tell you.

Ducks will help you by eating some of your lawn but don't let them in your mother's kitchen garden or oh, boy!

If she sees them she will come flying out of the kitchen flapping her apron and swinging her long handled broom and hooting to beat the band.

You don't want that to happen.

3. Pigs and Fences:

Before you ever start raising pigs in Big Lake, Minnesota you had better learn about fences or you will lose your pigs and everything else in sight.

You will want to have what is called pig fence or the pigs will wiggle out.

And you must have posts that you can put deep into the ground with a trench in between them to bury boards between the posts.

Dig your trench about a foot deep which should be enough or they will do what is known to pigs as rooting and get out underneath.

It is a good thing that pigs aren't good jumpers because pig fences are only about as high as you are or less.

And you must nail the bottom of the fence to the top of the boards you bury.

Your gate has to be very strong with a big hook in it to keep it fastened from the outside because pigs can find eighty five ways to open a gate.

When you have got your fence and gate done you will still need a pig house for real hot or cold days and some shade like a shed roof.

Pigs are very clean but their bath water gets dirty from when they get it and snort around and blow bubbles.

If you have enough pigs they will be called hogs.

To feed a pig is easy as long as it is wet.

If you have table scraps, potato peelings, dead bread or biscuits or anything else from the house you will put it in a slop pail for your father to carry out to them for appetizers.

He will dump it over the fence into their trough.

Then he will go to the barn and get dry pig feed and mix it with water and pour that into the trough too.

Then he will get lots of ears of corn and toss that all over the pig yard and they will gobble them all up fast except the cobs.

To call a pig to feed him you must say sooeee like your Uncle Oscar.

By the time you can say sooee they will be by their trough anyhow so it isn't as important as it sounds.

If you don't look for holes under the pig fence boards every day and fill them in with rocks a pig will escape into your garden if it is summer.

Then it takes everybody and some neighbors to get the pigs back into the pig yard.

Almost all pigs die in November or get shipped to market where they will end up their days.

If you see your father and some other men out by the barn with a barrel of boiling water and a rope over the branch of the tree beside the pump don't go out there.

How to Learn Music

While you are growing up in Big Lake, Minnesota you may as well learn how to play some music.

In most cases you will find that there is no other way to grow up alive.

Especially if your older brothers and sister are doing the same thing only harder.

They have had to do it too and have got a big head start on you and this is not much help because they will get smart with you and either tease you or laugh out loud at your mistakes.

But you should set your goals in life early and one of them is to get out of being a piano player.

This takes a good deal of time and some practice.

Most of your practice will be at whining about why do you have to practice or that you are late already for football, baseball, basketball, hockey, skiing, swimming, sliding, chores or a meeting you forgot at school.

None of these things count as your mother thinks that all of those things put silly notions in your head. She also knows exactly how much practice it takes because your sister who is going to teach you is what is known as an accomplished pianist who will some-day be world famous like Eddie Duchin and go on to MacPhail Music Academy in Minneapolis where only a few of the very best get in and she as already been recommended by another world famous piano player, Irene Jacobson, and has been accepted even though she is not even hardly in high school say nothing of graduated.

Anyhow whine.

Here is what to whine about.

None of the other boys in the second grade have to play the piano.

Piano players are sissies.

Your hands are too small.

Your feet don't reach the pedals.

You can't read music.

It is hard on your ears.

There is no future in it since you are going to be a cowboy and there are no pianos on the ranches where you will be herding little dogies.

Your friends will tease you.

Your brothers don't play the piano even when they know how.

Your father always goes out when you sit down and if he doesn't have to listen to you you shouldn't either.

Now, learn these answers and think about other things to whine about because these don't work.

The answers your mother will give you are:

The other boys don't have a piano, that's why.

Don't you ever say that about anyone. Your hands will grow.

So will your legs.

That's what you're here for.

It wouldn't be if you'd practice.

What in the world kind of an excuse is that?

Your friends are just jealous.

Your brothers have a firm foundation in music from playing the piano and chose the slide trombone and the coronet so they could play in the school band.

Keith is the one who plays the trombone in the band in his white pants and he says he puts banana oil on his slide to make it go real smooth. Keith is hard to believe sometimes when he smiles at you and I never heard of a banana well that gives banana oil, but he says he gets it from August Peterson in the drug store. You just don't know, do you?

If your mother doesn't give you these answers your sister will be able to because she has heard everything from your older brothers before.

No matter what, you are going to learn the piano and you are going to practice until you can play it or you will never leave this house on Saturday morning until you can.

You will soon be convinced that if you ever want to play an instrument you must get through the piano part first.

It all looks so easy, too, when your sister plays because she has accomplished.

She will be very understanding with you and can out-wait you every time.

There will be scales which begin at the middle of the piano called C.

A scale is what you start to play on the piano, both left handed and right handed and then together.

Your hands won't reach and neither will your feet but like they told you, you will grow and you must make do with what you have got which isn't much to start with.

But your sister will say that's it, that's it, good work and smile and hug you so keep it up.

When you have learned the scales by heart the fun
is over for a long time but listen to your sister and
repeat after her.

Play A and say it.
Play C and say it.
Play F and say it.
All in tune if you can like do re me fa sol la ti do only
whatever she says to play. You have to do what they
call in-pitch and is very different than what you think
it is even though baseball and kittenball may be on
your mind it is the last thing on your sister's mind so
forget that part and stick to the music. Pretend you
don't understand and then she will show you and
sing it out good and loud so everyone can hear it
plainly. Pretty soon she will do what is called accom-
panying you so you are both playing the same notes
and singing them too.

You will soon notice that there is no sign of your
cat or your dog. Max has left the steps for the barn,
and your mother has closed the door between the
kitchen and the dining room.

That is very discouraging, but your sister will say
good, good, very good and hug you some more.

When you get to the part where you have to do all
this on your own look out because your sister will
have prepared a fatal trap from which you may never
escape.

She will teach you Country Gardens and also The
Bells of St. Marys.

She will also say how would you like to learn this
one and play some very ritzy stuff without ever

looking at the music except to turn the page because she is all the time watching you.

Then comes the real music books and these are called Etudes and nobody can play that stuff even after a hundred years because they always get harder and there is always six million Etudes ahead of you.

The worst part is not the lessons from your sister it is the practicing every day after school for 30 minutes or one half hour.

You are all alone and the door between the kitchen and the dining room is wide open.

Your mother will say when are you going to start because I'm not setting this alarm until I hear you get serious.

You will learn to say:

I can't find my lesson book.

Sister never told me what to practice.

I have to go to the toilet.

Where is my pillow, please?

How long do I have to practice?

It's too hot today.

It's too cold in here.

The dog ate my book.

Oh, look, there's Mrs. Lannon going by.

What kind of birds are those on the clothesline?

What's for supper?

Can I help you in the kitchen?

Where is the metronome?

I think the metronome is busted.

Show me how to wind the metronome again, please.

Is it time for Jack Armstrong yet?

I'm not feeling good.

Can we go over to Uncle Doc's this weekend?

The answer to all those questions is the same.
It is:

Stop all the nonsense and practice.

Do not give more than two or three comments or questions a day or your mother will come out of the kitchen wiping her hands on her apron and her nostril will flare.

This is the danger sign if it is not already too late.

Try to estimate beforehand how much you mother is going to stand for before you start your whining, comments or questions.

It is usually not very much so be cautious.

Attack you lessons boldly with lots of enthusiasm.

Enthusiasm makes up for a lot of intentional or unintentional mistakes.

It also allows for some daydreaming if you have a good imagination.

Enthusiasm is impossible to keep up for five minutes, so you are pretty much stuck with at least 25 minutes of practice.

Do not stray too far off from your lessons because you mother knows these things by heart because she has been out there in the kitchen listening to the same mistakes for about 10 years.

She is an expert on mistakes, day-dreaming, scales, long pauses and making stuff up in place of real practicing.

At some point in your musical career you will

have made a decision to try something besides the piano.

If you like march music you will soon see the necessity.

There is no such thing as a piano marching band.

Stay away from reed instruments which could be clarinets, saxophones and so forth.

They are squeaky and the reeds get all chewed up.

Consider the coronet or the French horn or the alto but you are also probably still too small for the French horn but try it anyhow as it is very beautiful and old sounding.

Forget the tubas, too, because you will find them not only way too big but also they fill up full of spit from all the ompapahing.

Trombones and coronets and some others do too, but they have got what you call spit valves on them.

Never point you instrument at anyone while blowing out your spit valve as they will say oooogg and slough you one over the head with anything handy or their fists.

You will know when your spit valve needs to be blown out because your music will sound like it is bubbling or being played under water.

One thing about the band is that no matter how bad it is you will go along with the rest of the town when they go On to Nicollet.

On to Nicollet means everybody in town goes to a baseball game at Nicollet Park in Minneapolis to watch the Millers with Joe Hauser beat everybody except St. Paul.

The band goes in the two school buses and you get to Minneapolis in the afternoon and everybody meets at the West Hotel where men get free beer and you can watch them getting silly until their wives tell them to stop.

You also get plates of free beans, hot dogs, potato salad and ice cream as much as you can eat. Do not take too much or you will be sorry about the time you line up and march around the baseball field while you play Washington Post March and Under the Double Eagle.

It doesn't matter if you can't read music or do the fingering if you are a coronet player because you will just have to learn how to look out of the corner of your eye and do whatever the player next to you does.

You don't always blow it right but try anyhow and always quit when he or she does.

There are many girls who play clarinets, but not coronets because they think a clarinet or saxophone is more ladylike, whatever than means. Your cousin Donna Kiebel is not included because she goes to school in Monticello where she has become what is known as a coronet virtuoso, whatever that is.

Also, be sure you are watching and keeping in step most of the time and turning when they are turning or you could end up in the third base dug-out.

All of these things must be considered while you are learning the to play the piano because you must always have an escape route in mind or get stuck with about a hundred more years of lessons.

Huntington's Pasture

If your name is Hunt like mine you will wonder if Harry Huntington is related to you because he has a pasture of forty acres next to yours which is only three acres and his name is almost like yours.

He is not, and I don't know how much is forty acres except big.

Mr. Huntington has cows and horses and a beard so tough only Mr. Tschida the barber downtown can cut it and he doesn't want to because he will all the time be stroping his razor when he should be talking which is most normal for him with words nobody else knows.

Mr. Huntington lets kids loose in his pasture to do lots of things you couldn't do without trees and woods and ponds and willows and trenches and ruins from The Civil War.

And INDIAN MOUNDS.

Mr. Huntington says you should close any gate you open, do not fall off the stile and do not start any fires you can't put out.

Mr. Huntington does not care how much wading you do in his ponds or how much brush you use on projects as long as the brush is from his brush piles and not ones you have started yourself.

What you will want to do first is explore in the pasture with a friend like Thomas Ewing who couldn't get lost if you wanted him to and sometimes you wish he would.

Start out from a place in your pasture where you know where everything is.

Keep an eye out for cows and bulls and go the

other way.

Keep low and behind bushes until you get to the woods.

Do not go too deep into the woods the first time.

If you do you will get lost unless Thomas Ewing is with you.

Scout around the edges and make trail signs to follow if you happen to come back the same way, but you won't so don't worry about it because there is always Thomas.

As you get across the first clearing between your pasture and the woods you should notice that there is a pond down below.

Hope the cows aren't there with their boy friends.

Head down the hill which you will ski down in winter and get to the pond.

There are five billion frogs and no fish.

But there are water spiders and hornets and bumblebees.

And pretty flowers and plants along the edge.

You will notice that there is grass growing underwater and this must be explored to find out where the grass stops and the mud begins.

Do this by wading and yelping about how cold the water is, especially Thomas Ewing because he has got more feet to get cold.

On one end of the pond you will see that there is a huge old willow tree laying on its side and still growing.

I don't know how it does that especially since you can see that it has got burnt wood inside where it got struck by lightning as they say.

You can walk on top of it and get way out in the pond where it is probably deep.

Jump up and down and try to make the other kids fall off.

In the summer it is very nice there and soft and grows moss to lay on while you dry off.

In the winter time bring along a bunch of friends with shovels to clear the snow off to slide on or skate if you have them.

Best of all is to play hockey.

Hockey is a game where you choose up sides and use hockey sticks to make a goal with a hockey puck.

The goals where you score are small to larger logs and you must make the smallest kid lay down on the ice as the goalkeeper between the logs.

A hockey stick is a willow branch with a good bend in it or a dead broom stick with a board nailed to one end. You will soon learn that it is hard to pound nails through the board and into the round boomstick, but life is like that, my dad says. Two boards are even better as it will be stronger and maybe stay together longer.

A good deal of this depends on how much time you have to get ready.

A hockey puck is an old rubber heel which you can get lots of from Mr. McGintey's kids as he is a shoemaker and these are just old ones he has re-moved in order to put on new ones.

Of course you know the McGintely kids will want to run the game because they got all the pucks, but that is okay, let them.

If you can, get Chink McGintey on your side and Lefty McPherson and that's about all you should need.

Before you start your goalie must be standing.

Somebody will drop the puck in the middle and everybody go after it.

Everyone will be yelling and pushing and shoving and swinging at the puck on your shins.

If you notice, kids with skates get to the puck faster and they are also easier to knock over.

Always pick up your goalie if he gets stomped or skated over.

Brush him off real good and tell him what good work he is doing down there on the ice.

Promise him anything if he is crying.

If you happen to be in the open, yell and holler for the puck.

It might come your way.

When it is all over you go over to the fire to dry off.

Hockey players sweat a lot.

Goalies will be shivering and wet.

Make sure your goalie is dry before you send him home.

If it is night and moonlight you can play then, too, but it is harder to see and there are more older kids who swear at you.

This is all stuff you need to know like if you want to keep matches dry you put them in an empty 20 gauge case and put an empty 12 gauge case over it. It will be waterproof.

If you don't believe me stick it in a wash pan full of water.

When you have explored the first pond enough you will move along to what is known as the middle pond.

There is not much you can do with this pond as it has no willow tree which has been struck by lightning.

But it has cat tails for in the fall when you can take some home to your mother as a present in case you are late.

There are some very big red oak trees nearby on the edge of the woods which you are going to make use of when you want to build a lean two.

Now you are ready to head over to the far pond which is closes to where you are going to discover The Civil War trenches and an old orchard with INDIAN MOUNDS in it.

You can just forget about the far pond because there isn't much except for cows wading in it.

Cows will never attack you as long as they are wading.

What you do is get a whole gang of kids together and go play in the trenches and there is also a German machine-gun nest left over.

A good thing to have is some dried horse banure which must be good and dry or you won't want to pick any up.

For hand grenades.

If you have a brother with a whistle he will probably be the commander but that does not mean you cannot distinguish yourself with heroic actions without him and his whistle.

About all he can do is to whistle and yell over the top which you do and then charge the enemy in the machine gun nest or as Mr. Arvid Johnson, who was in the Rainbow Division during the World War and was decorated for being distinguished, calls redoubts. I think that means holes.

They are really trenches where the enemy is hiding and throwing horse apples back at you. You must take it by yelling and charging despite the hundreds of wounds and all the times you get knocked down by shell fire.

Throwing your grenades is a good way to get the fight started.

You cannot use the INDIAN MOUNDS except to hide behind to attend to your wounds as they are sacred and not to be defiled.

INDIAN MOUNDS are where dead Indians are buried with their arrow heads.

Do not take anyone with you except Thomas Ewing when you first go to explore the woods.

There you will find berries which you can eat called gooseberries which are really sour. And wild strawberries and some others with deadly poisons.

In the clearings there are some bushes that look like branches from pine trees but they aren't.

Keep those in mind.

When you get to a place you like, you can start your lean two.

Find some dead branches and chop off most of the limbs.

Prop them up against your big oak tree or against the tree branches.

Bring brush and pile it up around there so you have got as much space around the tree trunk as you need.

Make sure your tree does not have a spunk hole in it with spunky water or you will die from the stink.

Now put leafy stuff like soomack in the brush you have piled against your poles.

This is called camel flodging because if you were in it nobody could see it or you for the camel flodging.

Be sure to build another lean two by the middle pond against the red oak tree because you will need it as a resting place after being in the trenches all day or nearly all day with nothing to eat but your hard tacks which is soda crackers.

You cannot have too many lean-twos and they can also be used as teepees but don't build a fire in the center or you will burn everything up.

Build your fire outside the doorway after you have made a bare, bare spot with rocks around it.

Build your fire in the shape of a teepee and keep your fire small with no smoke or you will be found by the other enemy Indians or cowboys galavanting wild in the woods and you will be tortured or tied to a tree and left to die of wilderness.

Keep up the appearance of your lean-twos and you can use them while skiing in the winter.

Store your fire wood there, too, and plenty of it.

Skiing is done on long things called skis which have a strap in the middle where you stick your feet.

The one end is bent up and that is the end you want to be looking at when you ski.

Push your skis forward one at a time like you do with your feet on the living room rug when you want to give your brother or the cat a shock.

Use a stick in each hand to help you balance and also to give yourself a push.

Keep from getting your skis under brush.

Practice going down small hills first like the one by the first pond.

You will know when you are ready to go to the big hill down to Big Lake by the Pavilion, but be ready to sit down quick if you have to.

Tree Houses, Dug-Outs
and
Lean-Twos

Glossary of terms:

Hamber is Big Lake for hammer.
Chimbley is a common term for chimney.
Zink is Johnsonese for sink.

Picking out a tree to build a tree house in is hard work. So is building it.

First of all, go out in the woods if you have one. It doesn't matter how far you go out in the woods as long as you don't get lost because you are going to be dragging your building material out there with some help and it is no use having friends who won't work hard.

If you do not have your own woods you can almost always borrow a tree from your neighbor who will have one.

The most important part about a tree house is the limb you are going to start building on. Here is a list of things you will want to look for besides a limb.

1. A couple of hambers

2. A whole bunch of great big nails and a lot of smaller ones, too.

3. A saw which is not so sharp that you can wreck it.

4. Some old doors

5. A pile of boards with some lathes in it

6. Tar paper

7. Rope or your mother's clothes line

8. A flat rock to straighten your nails on with your hamber

9. If you don't know how to straighten nails,

practice

10. Be careful of your fingers

Shinny up your tree and look it over. Check it for branches which you can hang on to while you are hambering and sawing.

Be sure you are higher up than girls are likely to climb unless you want to get girls up there.

When you climb down, go get Corky Shultz to help you.

You will want Corky or somebody else whose father runs a lumber yard.

Go get your doors from the nearest source which is a deserted house.

Go there in the daytime and take your wagon or both your wagons.

Get your doors out to the tree, but don't bother grown-ups to help or they will cuss you and wonder how you got doors and where.

Now is when you find out whether your door is too heavy or won't fit the limb.

If your door is just right and so is the limb then put it up there.

Don't try to get it too level or nothing will fit and the rain water will not run out and your rug will stay wet.

Brace her up real good with boards and big nails.

Hold your hamber on the end of the handle like a carpenter is supposed to do.

If the door is not available or does not work for you because it is too heavy or does not fit, keep it anyhow. Someday you will find a use for it.

In Big Lake a person cannot have too many doors.

Sometimes you may be able to get long boards off a fence, like the ones around pig yards.

In that case use these and hamber the door to the fence post to replace the boards you have borrowed in order to get your flooring job done.

It is not always possible to get permission to use fence boards.

In that case you must rely on whining.

After you have got your floor up you will want to have Corky tie boards for the side and roof to your clothesline rope and not too many in case the rope breaks and you could smash Corky and there would be no help at all.

Put your roof at a good angle away from the place you crawl in or you will drown before you ever get to sit inside and hear the rain.

Use tin for the roof if you can get it anywhere and that will make the rain sound best in your tree house.

Rain on the tin roof makes you hum real loud and sing.

The tar paper goes over the cardboard to help disguise your location and purpose.

Make spy holes wherever necessary.

Nail some short boards to the tree for a ladder if you want to get girls up there.

Use a rope with knots in it if you don't want girls up there and hang the rope through the floor with a hole in it.

Some girls can climb a rope better than Corky.

These are good ones to have on your side if you

ever get invaded.

Go up in the attic or storeroom and find a rug.

Take it downstairs and ask your mother if you can use it.

She will say that you can.

Put the rug on the floor of your tree house.

Now lay down on it.

If it is good and hot you will smell the rug and it will smell good.

Tell Corky he was a big help.

Ask him to bring some more of his father's cigarettes.

He will do it because the grape vines taste terrible.

Pledge Corky to secrecy about your tree house even if everybody knows where it is because it will make you both feel better.

Talk privately about stuff.

Keep an eye peeled.

Wait for it to rain.

Dug-Outs and Stoves

When you are old enough to make a long handled shovel work you are old enough to build a dug-out in your pasture.

This is when you can use Corky and his father real good.

Pick a spot which is least likely to have big roots, but bring an axe anyhow just in case there are some stray ones and there will be.

You will probably want more space than you end

up with.

Dig down 'til you hit nice sand. It is called sugar sand but it doesn't taste like it. Put some of this sand in a pile to one side.

If you have got cows or horses in your pasture be sure to put up some snow fence or barber's wire around your dug-out hole to keep them from falling in and hurting themselves as well as wrecking all your plans.

Now go down to the lumber yard that Corky's Dad works at and sweep the cement warehouse. Fill your wagons after you have put the cement you swept up in cement bags.

Keep doing this until you have plenty of cement hauled to the barn where it will stay dry.

Do not try to feed your rabbits this as they will only sneeze and thump and so will you.

Sometime along the line Corky's Dad will tell you to go ahead and take any bags that have been busted and only have a little bit in them.

When you are sure you have got enough cement you will want to be sure you also have enough rocks to line your dug-out with or you will never sweep enough cement to finish all at once and you should.

Your dug-out will now be ten feet long and six feet wide and one end should have a two foot place for your stove.

On one side at the end you will want an incline for your doorway.

Be real sure you have every inch of the floor lined with rocks and small stones.

Mix your cement in a tub, four shovels and or maybe five to one shovels of cement. Pick out the biggest chunks of weed from the floor of the cement warehouse. Throw in some small gravel which you have also taken from your dug out but no dirt.

Dirt doesn't make cement work good. Have Corky add water from the barn pump while you mix it down in the dug out with your father's garden hoe. The old one.

If your Corky is real strong, let him do the mixing as it is tiresome work.

Pour the cement on the rocks and smooth it out with the trowels you have borrowed from Corky's father.

I forgot.

Before you mix your cement you want to put post holes down every three feet and fill them with posts.

On the end where your stove will be you must put them at two feet or you will have covered up your stove place.

Make sure your fence posts are real straight and that some gravel and cement get into the holes after the posts are in so they stay put because you will be using them later.

Now do the floor and hope it doesn't rain.

Make your floor real smooth because you want it to be real nice so girls will like it if you can ever get one down in there.

When you get to the entrance you can go up the incline and haul everything out with you like your tub and buckets and trowels and gunny sacks you used for padding for your knees on the rocks while you were smoothing out the cement floor.

The next day you will want to put two more posts in but they should be at the middle of lengthwise.

They should be two feet back from the edge of your dug-out and stick up three feet when you're done.

Use your mother's yardstick.

Three feet and a yard are the same thing.

Bring the little two-lid wood stove you found at the dump and have it ready for you and Corky to let down into the two by two stove hole with strong rope from the barn. Get a very strong two- by-four ten feet long from somewhere and nail it to the top of the two posts. It will just fit.

Your cement floor has to be good and dry for two days before you tie your ropes to the stove so you can lower it in the stove hole.

Put tin around the back of your stove hole to keep dirt from falling in and it will also push the heat into the dug-out.

I don't know why, but it will.

Now, take all the doors and long boards you have on hand and make your roof, but first put Corky down in your dug-out with a broom handle.

That is the way to lower doors from the ends to the ten foot two by four.

Before you put anything leaning from the stove hole take an iron fence post and lay it down across the opening so you will have something to hold up your doors besides dirt.

Underneath the long two-by-four put some kind of wooden post or brace at the middle. This will interfere with some activities when you get done but it will be a good thing to have to hang your stuff from and to dry your clothes, too.

Now you can put the tar paper over your roof and tin with a hole in it over your stove hole for you stove pipe.

You must shove the stove pipe up from underneath because one end is sort of mashed to fit your chimbley hole on top of your two lid stove from the dump.

If your stove pipe has got what they call a damper in it, that is very good.

If it doesn't, don't worry about it.

You should build your steps going down like a ladder and then just brace the bottom end on the edge of your nice cement floor.

Make a door and some sides for the entrance as best you can to keep out the snow and wet.

You will need cardboard from the lumber yard which is from doors and windows that come in there from somewhere for Corky's Dad to sell.

He will give you all you need and then some.

Take some narrow boards and nail them to the top of your posts that you have all around your dug-out and set in gravel and cement as you remember.

Now take your cardboard down and use the side that doesn't have any words on it and tack it up for walls.

It will be just beautiful.

Now be sure you have a hole in the wall somewhere handy that you can hide all your valuables in.
Put a small wooden box in there.

Like cigarettes.

And secret messages and codes.

Go and get some sturdy boxes to sit on or some
chairs and a wash bench unless your mother says no.

Any of them will work.

Ask your mother and dad to come look-see. They
will say nice things to you and grin at you. Don't let
them look too close or your secrets will be out.

Go down in your dug out on a sunny day and sit
in the dark.

When you have done that for a while you will
begin to see places you missed in the roof and sur-
rounding area.

Tell Corky to patch them real good.

And stick a small object in the holes so he can find
where to patch. Say that's it, that's it, to help him
along.

Take some of the dirt you have left over and put it
all around to keep out the snow drifts.

I forgot to tell you how to board up the sides
where the roof comes together with whatever you
have on hand.

As you will see soon enough there will now be
very nice ledges inside your dug-out on which to
store your supplies and provisions.

Everything will get musty.

Especially bread.

When it starts to get hot out you will be nice and
cool.

When it rains you will be dry to start with.

When you start going to your dug-out you will want a kerosene lantern or two.

They give off much needed heat in the winter.

Make Corky bring them as yours are in use in the barn.

Your Dad has plenty of kerosene in the barn in a brown barrel.

Don't spill it.

Load your lanterns out in the sunlight.

Never start your fire with kerosene unless you have to, but start one while it is still warm and you don't need it so you will know that it works good.

Get a lot of wood cut up and store that on your ledges inside while it is dry and stays that way. You can get nice kindling from the room at the lumber yard that Dutch Maas uses to cut the ends off boards and 2 x 4's and like that. Dutch says he never makes a mistake, but he has got parts of his thumbs missing and Corky's dad says that didn't happen when he was a baseball catcher and a good one.

Bring our old cooking utensils and spoons and pie pans.

Practice cooking and eating.

When you are there after the first snow and it is blowing and drifting outside your dug-out you will know what snug means.

And warm and dry while the wind is whistling 40 per.

You and Corky will smile and laugh.

It helps if your Corky is red-headed because they are real lucky.

How to Go Camping

Before you decide to go out camping there are several things you need to know.

Where to go.

How to get there.

Who to go with.

What to take along.

When and for how long.

Things to do.

Where to Go:

As you know, there are many places around Big Lake, Minnesota that are bound to be wonderful for camping.

It is very difficult to choose only one place because there are so many and so are the ones on your Uncle Oscar's farm.

The place you choose will be a good one and you can always plan to go some other place later.

Or even next year.

So why not pick a spot where you know there will be fish because you must live off the land and water if you are going to survive.

That means one of many places because there are lots of fish in the rivers or lakes.

You must think hard about the Elk River, the St. Francis River, the Mississippi River, Bertram Lake, Long Lake or Birch Lake.

You will want to see my notes on those places but you should pick Elk River for the first time because it is just as good as any of the others and better than some.

How to Get There:

Getting to the Elk River is real easy.

Getting all your stuff there is not so easy.

So plan to get your Dad to help you which he will gladly do because he will wish he was going with you.

How to get there has a lot to do with where to go.

Before you get this far you should make sure you can go.

Then you will have to make repairs on your bike and find an air pump.

It is in the trunk of your Dad's car.

You will need your bike to come back to town with or to make surprise raids with while you are camping.

Mostly you will do that on foot because the Indians could cut across and get you and you would never know what hit you as they are so fast on their feet and can run all day if they are chasing you or are carrying secret messages from one camp to another and will only hand them over to the chief of the tribe.

Anyhow, draw your map so that you go out the Eagle Lake Road as far as the dump and then cut down the high bank into the woods at the big bend in the river.

Your father will leave your camping stuff at the high bank.

Leave your bike there too.

You will have to make many trips up and down the high bank out there as there is no other way to do it so make the best of it and make sure everybody else does, too.

It is very beautiful down there and there are many huge trees along the river which are red oak, maple, ash, birch and poplar.

There are many squirrels which are gray and a few which are red.

There is also a black squirrel which is very unusual since the rest aren't.

Black squirrels are messengers of death and you had better hope they are headed the other way when you go camping.

Who to Go With:

Choosing camping partners is very easy to do.

They are all your friends.

Some can go and some can't.

Usually that leaves you with from two to four.

One won't have a bike so double up and take turns pumping.

If your are lucky one of the friends who can go has got a hatchet.

More on that later.

When you have found out who can go and who can't you will also hope that one is John Wigren, one is Corky Schultz, one is James Gahr and another James, James Young.

At least one of them can go but has to be home nights which is okay because the tent is too small.

You can always prop up blankets with sticks for more room but if it rains oh boy!

And it will.

You have about all you can keep track of as far as

friends are concerned, but you will also learn who your real friend is because he will be the one you are hanging around with while the others are off some place talking about you.

It is also very good to have that one person who can't stay overnight James Gahr because he is rich and can bring back many emergency supplies like store bread, cookies, candy, chewing gum and cigarettes or pipe tobacco.

Pipe tobacco is best and then you can roll your own and always have dry cigarettes which is more than can be said for ordinary Sensations if you drop them in a puddle.

Be careful about inhaling.

Pipe tobacco will make you sick if you inhale it too much but you'll get used to it or quit.

When all of your friends who are going camping with you are ready to go you must get together and study the map.

Especially James Gahr who will probably be making the trip several times with emergency supplies and needs to know exactly how to get there and back.

Your Dad will already know how to get your camping stuff out to the high bank but be sure everybody is going to the same place even if all of you are going together.

Always practice.

What to Take Along:

After you have decided where to go, how to get there and who can go you still must find out what to take along.

Sometimes you already know who can bring what because you know they have it.

That has nothing to do with groceries.

Nobody has groceries. Only parents have groceries and they can be home made or from Kolbinger and Draves.

You have to have a good memory or be able to write lists.

Most lists will only have one or two things on them so I guess they are not really lists but it is good practice for you to be ready to make lists as you grow older.

You need to have:

A tent
Frying pan
Blankets
Clothes
Fishing rod and reel and line with hooks and spoons.
Rope
Root Beer
Knife, fork, spoon
Pie pan full of pie to start with
Matches
John Wigren's hatchet
A pot

For Groceries you will want to take:

Bacon and some meat
Baked Beans
Bread
Butter

Flour
Salt and pepper
Cookies
Root Beer
Pickles
An apple
Eggs in a jar
Jelly beans
More baked beans
Fresh tomatoes and be careful of them
Syrup
Potatoes

When to Go and for How Long:

You should go camping whenever you can if it is when school is still going on and it is only for Friday night, Saturday and part of Sunday.

If you think you are going to get out of a bath by going on a week end you are crazy because your mother won't need the smell of smoke to tell her you missed a bath. You will take your bath Sunday no matter what.

Missing Sunday School is not a good practice and you will only get away with it once.

But summer time after your garden is past the weeding stage is best as you can imagine.

Plan your camping for a week but you may only last four or five days.

Some of your camping partners will only be able to stay three days before they think of reasons why they have to go home.

Stay as long as James Gahr keeps coming back.

He will get tired of it soon enough so don't worry about being there all summer.

You folks will settle it anyhow.

Things to Do While Camping:

There are many things you have to do before you can do anything else.

Such as getting all your equipment and groceries down to where you want to camp from the high bank.

The best way is to put a lot of things in a blanket and four of you each take a corner and start down hill.

If there are only three of you let the other two do it.

Let the tent roll down by itself until it gets tangled up in the brush.

Then kick and push it.

It will get there but it may have some dirt and weeds on it and to get it to move on the level part you will want to tie your rope to it and pull it the rest of the way.

When you have finally got everything there try to find the hatchet.

You will need it to cut some poles for the tent and some short sticks to tie the bottom to.

The tent poles and stakes were lost years ago.

When you have got the tent up you want to put your blankets down for your beds.

First you put a piece of canvas on the floor.

Then dig a trench all around it.

Any rocks you didn't see before will show up at the time you first lay down.

Lay down now and find them or you will be up half the night.

Next you will have to do two things:

1. Dig a fire hole.

2. Send out a scouting party with the hatchet for fire wood.

Your fire hole should be started before they leave with the hatchet since you have to slash up the ground so you can dig. You will wish you had put down shovel.

Now go down to the river and pick up some rocks about six inches around or any good sized flat rocks.

Take these back and line your fire hole.

It will not take long for whoever went to get fire wood to get a whole bunch because there is so much.

Put a few bottles of root beer in the river to cool.

Also get two green branches with forks so you can put them on each side of the fire hole with a stout green stick between them.

This will be where you will hang your cooking pot.

Eventually your stout green stick will burn and the pot will end up in the fire unless you are real careful in which case you will not lose your food if there is some cooking when the accident happens.

You will soon discover it is much better to fry everything in a long handled frying pan or change the stout green stick every day.

Everyone will want to put their groceries together to see how much you have got but first you want to start a fire even if it is 100 degrees above zero.

To do this tell everyone to forget using paper because you can start a fire every time with only one farmer match.

Take out your red jack knife and whittle on some nice dry wood to make it real fuzzy.

Do this to a lot of sticks and place them in the shape of an Indian teepee in your fire hole.

Light your match close to the dry wood and start it burning all around the bottom.

Now add some more wood but be sure to keep in the shape of a teepee

Add more wood.

It will have gone out so crumple up some of your newspaper and start over.

When the fire is going good you will want to get everything ready for supper.

This means decide whether to go swimming or fishing first.

Everybody will probably want to go swimming, but that is not a good plan.

You will scare all the fish off.

If somebody has to go swimming he must go around the bend downstream.

Fish cannot hear around corners.

So now you can wade out in the river or just cast from shore.

You will catch some nice Northerns which you must somehow clean.

James Young is very good at this and can have a fish cleaned up in a short while and therefore ready for the frying pan.

Take out you pie pans and put bread on them. Spread some of the butter that melted on the bread.

Then heat up your beans and put them in the hole by the fire to keep them warm.

Open your home-made pickles.

Put a whole bunch more melted butter in your frying pan and let her sizzle real good.

Put flour all over your fish pieces and some salt and pepper, too.

Take your pieces of fish and put them in the frying pan.

You never tasted anything so good.

Be careful of the bones.

Don't forget the tomatoes

Drink some root beer and burp.

You will have to take turns washing the frying pan and pot in the river.

Scour them clean with the white sand on the bank and rinse them in the river.

Wash your pie pan and fork the same way.

Lay around the camp fire and make up ghost stories or watch the moon and stars or both.

It will soon be time to go to bed and you will sleep real good.

Especially if it rains and there are no leaks in your tent and your drainage ditch works.

No matter how hot it has been you will need at least two cotton blankets.

Wool blankets itch too much but James Young likes them anyhow.

He says that is what they have in the Army and it

shows you are tough and a good soldier.

You will hear the rain on the tent and see the lightning flashing and the thunder will be very loud.

You will snuggle down which is wonderful and slowly go to sleep and stay that way 'til after the sun rises.

In the morning the first one who has to get up must build the fire.

If you are smart campers you will have put some firewood inside the tent to keep it dry.

Be sure to brush you teeth so you don't have to lie to your mother when you get home.

Then you should gather some more wood and argue about breakfast.

You will end up making some pancakes with your bacon.

If you don't know how to make pancakes you shouldn't be camping. Ask your mother.

Don't hog all the syrup.

Check the syrup level everyday and if it is getting low tell James Gahr. He will bring Log Cabin Imitation Maple Syrup and that is the best, but Karo Corn Syrup is also good.

After you do your dishes take your blanket outside to air. They will need it.

Don't do it if it is raining.

Hang them over a bush.

There are lots of things to do besides swimming and fishing.

You can go for hikes and look for new places to fish or swim.

You will have a lot of fun after a while you will be ready to do it again once you have gotten home.

Before you go home you must do what is called breaking camp. This means that you have to get ready to haul everything that is left back up the high bank. You must also dig a big hole and bury all of your garbage except left-over food. Take all your stale or dry or moldy bread and break it up and then take it to your best fishing spot. When you toss it in you will be doing what is called seeding the river because the fish that eat it will hang around expecting more and when you go fishing there again they will bite on your spoon or bait. My dad says the fish in Elk River aren't very smart that way and will lay there for years just waiting to be fed.

I don't think I should believe that, but that is what my dad says.

Talk to your friends in a friendly way and stay on the good side of your folks.

How to Get Ready
For Christmas

It is never too early to get ready for Christmas and here is how to go about it in Big Lake, Minnesota.

You will get getting your Fall and Winter catalogs from Monkey Wards, Sears, and Spiegels.

There is nothing in Spiegel's that you can't get a better deal on from Monkey Wards or even Sears which is not as good on clothes but is better on tools, especially.

You will also be getting special sales catalogs from those companies so be sure to keep them all together.

Do not wait for the sales catalogs before you start making your plans on what you should get lined up for your folks and your brothers and sister if you have got any money or are going to earn some by doing jobs of work.

Get yourself a nice lined Big 5 tablet if you can. This will help you to get organized and do some writing which will be mostly copying because of the big words.

You can use a pencil or crayon.

Put Dad on one page and Mother on another and then on another page put your sister's name and then some lines and then your oldest brother's name and then some lines and then your other brother's name.

If you have got more brothers and sisters than that, just keep going.

You will soon realize you aren't going to have enough money.

That is why you must get everything in order so you can make decisions. If you don't know what a decision is you probably are too young to have any money, but if you do have money then a decision is a really good guess.

When you have got your pages you need to start studying the catalogs. You will already have some good ideas of what your brothers and sister want because you have heard the same story over and over.

With your folks you will be able to really surprise them because they never want anything and will tell you not to waste your money on them because they have everything they need.

And will tousle your head and smile at you.

Here are some good ideas for looking in the catalogs and copying the words you will need.

Mother:

    Beautiful textured warm wool coat,
    fully lined in satin with domestic fur collar
    and matching cuffs, raglan sleeves
    100% wool, colors: Baby blue, Navy blue, Tan
    or Gray. See hats to match page 386
    Wt. 4 lbs. 12 oz. Sizes 8-28 Price $19.60

    This dress comes from the Chicago store.
    Please allow seven extra days for delivery.

    Chic rayon party dress with unique
    mandarin collar which folds down
    if jewels are to be displayed. Belt
    in contrasting red suede fabric.
    Sleeves on long sleeved style may
    be worn full length or pushed up

Wt. 4 oz. Colors: Dk Br.
Tan, Fedora Gray
Price $3.20

See Hat blocks page 722

Canadian style Mackinaw
made of 100% wool with double
stitched leather buttons
Your choice of plaids, red,
yellow, green, blue. Lined
with combination wool body
and cotton sleeve. Big, comfortable
pockets for your every need.
Chest sizes 34-48 Sleeve length
28-34. State both when ordering
Wt. 2 lb 6 oz.
Price $8.25

Chinese split bamboo hand made 14 foot
fly rod in its own leather case.
Four clear laquered tapered sections
with slip in no lock seating. Handle
accepts any style reel of similar
quality.
Complimenting fly tying kit at
no extra cost.
Wt. includes case   2 lbs 10 oz
Price $12.40

Sister:

Ladies seventeen jewel Swiss action
watch. Pin it on your lapel or carry it
in your purse. 14 carat gold case and
beveled glass crystal. Five year
unconditional guaranty. Made in
Switzerland.
Wt. 1.5 oz
Price $9.50

Electric hair curling iron.
Plugs in any conventional
110 volt outlet. Ready to
beautify your hair in only
four minutes. Safety assured,
no over-heating, no cleaning
necesssary as with lamp curlers.
Ideal for today's traveling
woman. Modern hairstyles
booklet free with every Sears
Lady Diane.
Wt. 5 0z.
Price $1.50

Oldest Brother:

All leather high top basketball shoes
with non-marking black rubber
soles and arch supports. Our
best sports shoe. Comfortable

and made for many seasons of
court wear. You'll score high
with this special bargain.
Black only. Sizes 6 1/2 - 11 also
in half sizes D-AAA
Wt. 1 lb 1 oz.
Price $2.25

Fox Savage 12 gauge Double Barrel Shotgun
2 1/4 in. chamber
Right barrel full and left barrel
modified. Barrel length 26 in.
Walnut stock with raised cheekplate
and soft rubber buttplate for near
total shock absorbtion. Order
before Christmas and receive
certificate for 2 boxes Peter's
hi-velocity shells with your choice
of shot size.
Shipping by rail only.
Wt. 5 lbs 6 oz.
Price $19.50

Conn's Silver Shadow trombone
as pictured in "A". Easy glide
action, elbow tuning and perfect
bell-tone. Velvet lined case with
polishing cloth and 1 oz bottle
banana oil for slide.
Shipping weight in case 3 lbs
9 oz.
Price $16.80

Other Brother:

Schwinn's Blue Streak Boy's
Bicycle. 26" or 28" model. Rubber
grip handle bars, New Departure
coaster brake, saddle leather seat,
front D battery light and safety
reflector. Rear view mirror and
streamlined chain guard. Balloon
tires. Puncture kit and air pump
attaches to frame bar between
steering post and drive gear housing.
Railroad Express only
Shipping weight   14 lbs
Price $20.50

Lionel HO Guage Train Set
Hiawatha model engine with
six lighted passenger cars,
caboose. All electric with
battery drive and 12' of HO
track, headlight whistle and
one set crossing lights. Large
selection of additional cars,
side track, depot and platform
on page 541
Wt. 2 lbs 3 0z
Price $9.50

rope and tinsel and more lights and it will snow after dark and you will see the tree as the snowflakes come down out of the sky.

If you open your mouth and hold it just right there will be some snowflakes dropping on your tongue.

And in your eyes and you can look out and it will be like a lace doily.

Be sure to go to Johnson's Hardware Store, too, as you will need to compare tool prices.

Then go to August Peterson's Drug Store and Ice Cream Parlor to see if there is anything there that you haven't seen before which could be used as a gift or is new.

August Peterson's is not a good Christmas place except for Christmas cards which you must make yourself so forget it.

He has got decorations everywhere and if you go to the back you will find the great big floor register where the coal furnace in the basement is red hot.

Stand there until you smell your overshoes burning.

You will be warm all over and ready to say goodbye thank you to Mr. Peterson or Mrs. Peterson or both and their nice daughter who is your sister's age and very nice because she likes your sister a lot and will talk seriously with you if you want to.

Make your gift purchases where you find the best gift items which will be Bub Garver's and Johnson's Hardware Store.

Here is what you will want to really buy.

Mother:
   A paisley scarf (square)
   Color purple
   Wt. ?
   Price
   25 ¢
Father:
   One package of new hack-saw blades
   No color but blue
   Wt. ?
   Price
   29 ¢
Sister:
   New Song book
   Color yellow
   Wt. ?
   Price
   15 ¢
Older Brother:
   Plaid shoe laces
   Color ?
   Wt. ?
   Price
   10 ¢
Other Brother:
   Spy glass from Japan
   Color: Brass
   Wt. ?
   Price
   10 ¢
   One large package of mixed color tissue paper and

a roll of ribbon:

Price

5 ¢ ea.

When you have got everybody's Christmas gift wrapped and tied and their names printed on your nice home-made Christmas cards sneak downstairs and put them under the tree.

Look around and see if there is anything there for you but don't touch or move or your mother will know.

Christmas Eve you will go to church where your sister will sing solos and everybody else is singing and they have a manger and baby Jesus is born on the front end of the alter and some of your friends are angels or have got sheets on the shepherds on a midnight clear.

When you get done with that you go home and eat cookies, cake and drink hot cocoa or Baker's semi-sweet chocolate melted in milk which is better if you can get some sugar into it.

Then you must go to bed after you say some special goodnights and try to sleep.

There is no use.

But later on you will anyhow unless your brother is kicking you which he does when he sleeps.

Then when your mother calls Merry Christmas up the staircase after opening the door for warm blast you will be quick to get down to the living room.

Your father will pass out the presents as he is the boss.

Here is what you will get.

A scarf your mother made.

Some sox

A pair of mittens with a string between.

Some nice handkerchiefs your mother sewed your initials into.

A spinning top you can spin on your dining room linoleum or in the kitchen if it isn't too busy.

You will also get something from your sister and each of your brothers which will surprise you because you can never be too sure how you stand with them.

Save your paper and ribbons.

Put them in the box your mother has got for just that purpose.

You will not even notice that you have not had your breakfast yet but don't worry because your father will fry some nice potatoes and you will have scrambled eggs and your bacon and home-made bread toast or morning glory biscuits and lots of sweet cream butter and homemade strawberry jam and homemade honey and seconds on everything.

Then you will want to go back in the living room and look over your brother's and sister's gifts and do some exclaiming about them and thank everybody all over.

In the afternoon you will go over to your Uncle Doc's or out to the farm to your Uncle Oscar's. Whichever place you go all your cousins will be there and you can tell each other about your presents and Grandpa Kiebel will set you on his lap and give you sugar cookies or some of his mouldy cheese which he makes himself but is good if you don't look at it or smell it too much.

The car will be very cold so you will need the big old buffalo robe from the storeoom, so remind your oldest brother to bring it along for the back seat especially on the way home after dark.

You will be good and happy.

And sleepy.

How to Mow Your Lawn

If you have got a house in Big Lake, Minnesota you are going to have to mow the lawn but you should rake it first. You will want the iron one that gets stuck in the roots and makes you mad and also the one which is wide and looks like a mashed flat broom and is called a broom rake.

When it is time to rake your lawn your mother will tell the whole family and it will get done even if your father has to go down town or fix the rabbit hutches in the barn.

You will also know who will be able to accomplish the lawn raking if you have older brothers and/or sisters.

It is not them.

However if you whine enough your sister will rake near the road where some people might go by. Your mother will do the hard part of the hedge and your father will fix the garden cart so it is easier for you to haul the leaves and branches and dead grass and animal remains which became remains after your dog got through with them.

Animal remains do not burn as good as they should in your bonfire so it is better to bury them, but not where Mr. Sullivan and his horse will plow them up when he and his horse whose name is Nell plow your garden and half the driveway because when Nell gets going she is hard to stop in the Spring.

Most people in Big Lake already know that.

You should also make sure you understand where you are to dump the grass and leaves and branches or twigs so you don't have to move it. Mothers and fathers seldom agree on the location any more than they will agree on whether it should be one big pile or lots of small ones.

If your father gets to decide he will have one big pile and he will be there to burn it for you at night with a five gallon bucket of water with a gunny sack in with the water to swat out anything in four countries he sets fire to.

If your mother has time she will decide to have you put the stuff in smaller piles and burn it that way because it is more ladylike and she won't worry about burning down the house and garage and barn and the woods and pastures clear out to the Mississippi River and back to the gravel pit even if the wind is the wrong way to do that in the first place. But she will still worry and won't let you do it yourself even if you have got a gunny sack and a bucket of water full to the rim.

The colder the water the better.

When you are burning after supper there will be many kids there unless they are burning too at their own house. They will form groups or make a circle around the fire and cannot hear their folks calling but that is okay because the neighbors like to come over to look at the fire and make remarks about the weather and their lawns and gardens and brag about things and same kids they were screaming at less than six minutes ago have done in their extraordinary lives in Big Lake.

Mr. Zetterberg will sharpen your mower real good and it will work very good on the way home as you test it along the railroad tracks and ditches and on the edge of the sidewalk but it should because he will charge you plenty, a quarter.

Your father could do it but he doesn't have the tools.

The lawn mower works best if it is a good color and has got rubber on the wheels and goes tickety-tickety-tickety when you push it real hard and then jerk it back so the rotary blade spins real fast.

Do not clear the blade of obstacles or sticks with you toes as you will be barefoot to get better traction.

Do not expect to start too early in the morning Saturdays if your neighbors don't get up 'til seven or so and besides the dew makes the grass slip through without being cut.

Another thing that will happen is that you will discover that the grass cuts higher on one side that the other and that is because on the way home from Zetterberg's you have damaged some rocks and nails while testing your mower.

Do not bother your father.

Slip into the garage and get your Dad's good screwdriver and move the mower to the other side of the hedge where you can work on it in peace.

Turn your cap backwards on your head like the race car mechanic or this won't work.

If the grass has been cutting high on the left side you want to loosen the front screw on the left front side of the bottom cutter bar which runs the width of the mower and is right behind the rotary blades which turn parallel to the ground.

Then tighten the screw on the cutting bar that is directly behind the one you just loosened.

No, that was wrong.

You will want to do just opposite.

Loosen the back one and tighten the front one otherwise it will be cutting even higher.

Then take your screwdriver and turn the screws on the right hand end of the cutter bar just opposite of what you did on the left.

Try it out and see if that doesn't fix it. If it doesn't try it six or seven more times.

Turn you cap back to the way you were told to wear it.

Watch the house as you take a screw driver back to the garage and put it back exactly where you found it. Make sure the grass in cuffs or your bibbers didn't leave a trail on the dirt in the garage where they can trace you.

Go in the kitchen as ask your mother where your Dad is. She will know.

Go where she says and ask your Dad if he will get his screwdriver and fix the mower. He will ask you why, etc., but will come out and ask what the mower is doing way out beyond the hedge and you will tell him you didn't want to disturb them.

Be careful around the flower beds and do not bang into the shrubs and small trees with the guard bar or you will hurt them.

After a while you will be able to hold one end of the mower up with the cross bar handle so you can do edges but you must be strong and determined and get some compliments or you will never be able to do it.

I don't know what you do about tiger lillies not to cut some of their green stuff.

Going around trees work best if you dig out the grass around them with a space and throw the grass and dirt in the road to get run over and flattened out so it disappears automatically.

Then you can mow right up to the tree and clank into it with the guard bar. Nobody cares because the grass is trimmed just right because you have done good work.

When you have been at it for two hours and fifteen minutes you should go in the back door and ask your mother if there is any lemonade left over and there is bound to be as well as two fresh ginger snap cookies which you will eat while sitting on the cellar door or on the yard swing where you will have just finished trying to mow underneath without getting killed alive.

As you sit there you will notice that the robins and the bluebirds are enjoying your fresh cut grass and tossing it around as well as doing a lot of head cocking and worm pulling and having some fun.

Do not take your lemonade glass back inside or your mother won't be able to ask what you did with it.

Keep an eye open for Bill Glenn and the ice truck as you will want some free ice slivers when he is filling your ice box. He will always let you have some because he will want some lemonade too and he will take it, too, as your mother always leaves a clean empty glass in the ice box so he can do that.

Or some homemade root beer which you cannot have in the morning but he can and always does if there is any in there previously opened or fresh.

When you are about fifteen minutes from finishing up your brother will come from somewhere and tell you he will help you.

Let him.

Go tell your Dad you are through.

He will ask you who is mowing if you aren't

Tell him.

He will know who did all the work.

He will thank you and look at you and smile.

Your mom will probably wink at you and grin as you go through the kitchen door.

How to Get Ready for School in Winter

When you are going to school in Big Lake, Minnesota you will need your mother to help you, but there is a lot you can and should do yourself.

It will help if you are organized.

And stay that way.

Here is how you do it in winter.

When your mother calls up the stairs from the kitchen that it's time to get up you will sure know it.

If you lay there for a few seconds you will smell coffee and bacon or pancakes through the cold air.

You cannot smell cocoa in cold air as quick which is for you and not the coffee.

Try to move your legs and slide out from under the blankets if it isn't cold enough for a feather quilt yet.

If it is, you are really in winter.

Try to stand on the register and reach for your long johns, black socks and other school clothes.

Your brother is probably already standing on the register as he has got rights.

Rub your eyes and say good morning to your brother so he doesn't get smart with you.

Grab all the clothes you think are yours and get downstairs as fast as humanly possible.

You will feel it getting warmer as you open the stair door and then it is very warm in the kitchen because the wood stove is going good and so is the furnace in the basement because you are burning red oak which is the very best your Dad has got to put in it.

For heat.

Red oak will have been put in the furnace the night before because it burns longer, too.

Your Dad knows all that stuff.

He will be in the kitchen putting on his boots and your Mother will be taking food and plates out of the warming oven and dipping hot water out of the resevoir for you to wash up with at the sink.

The Johnson boys call it zink.

There are only four seconds to get to the furnace register in the dining room before your oldest brother and sister get there.

Make the most of it.

They will soon be there to push you out of in front and into the chimbly corner where you will finish dressing as best you can and then wash up and brush your teeth.

And comb your hair.

And wipe your shoes.

I don't know why since they will soon be inside your overshoes anyhow.

Get to your place at the table and smile at your Mother.

Do not ask her what is for breakfast.

She will show you soon enough.

If you don't drink your cocoa quick it will cool and get some dead skin on top.

So drink it.

There is plenty more.

The homemade bread toast will hold about one half pound of butter per slice because it has got lots of air holes.

The bacon is from a pig.

Try not to think about that and put mustard on it like your Dad does.

It will taste twangy.

If you have pancakes as you suspected there will be even more butter used and syrup which is called Karo and comes in gallon cans.

White or dark.

Your eggs will be fried a lot.

That is the way it is.

Do not dawdle as you mother says you are doing.

She will seldom get to the table except to put more food on.

She will be preparing you lunch which is:

Two sandwiches of meat and catsup (home-made)

Some cheese (bought)

A Winesap apple (homemade)

A nice slice of pie (Apple, Cherry, Peach, homemade)

Milk (Mr. Swanson's)

She will put this into your lard can which is called a lunch bucket at school.

Before you get dressed for outside count your brothers and sister.

If one of them is missing do not head for the outhouse or you will freeze to death waiting.

Stand by the door and as soon as you hear the latch on the outside door open, get set to run and then run like crazy.

Always put down your lunch bucket before you

try to put on your cap, scarf, sweater, coat, mittens and overshoes.

When you have learned how to do all that and buckle your four-bucklers you will be ready for second grade.

Make sure the string on your mittens is behind your neck or you could strangle to death on the way.

Put down the ear flappers on your cap.

You will be told to wait for your brothers or to hurry up.

Someone will help you find your way to the back door.

If it has snowed and drifted you may be able to walk part way on top of the drifts.

If it doesn't work, keep on trying.

When you are part way you will see some other kids going to Big Lake School, too. Those will be your neighbors Joyce and Donna Larson and a whole slough full of Johnsons if they are going to school that day. Some days they have to stay home to help their father, Howard Johnson, with chores or loading cattle to go to the market in South Saint Paul, but I don't like to even think about that.

Your brothers and sister will have left you to freeze if you don't hurry up.

They never wait so make some angels or start run-fox-run on the vacant lot across from the church.

By the time you get to Hadden's corner you will have a minimum of three friends with you.

There is a lot to do before you are ready for school to start.

Snowball fights is one of them, but don't lose your lunch bucket.

Running and sliding is another.

You will find it difficult to keep your scarf up over your nose to keep from freezing it.

When the first bell rings make a run for it.

Bang your feet on the school house door to get most of the snow off them.

Stand inside, too, to get the ice melted off your four-bucklers or you won't get them undone.

Go to your room and say good morning to Miss Hoyt who will be standing in the door to her room and hugging herself because it is cold but she will smile at just you.

Go into the cloak room and put your lunch bucket on the shelf over your coat hook.

Take off your scarf and coat and mittens and if it is warm enough in there your sweater.

You will have taken off your cap and smoothed over your hair when you came in the door.

Take off your overshoes.

Your Mother says do not take off your overshoes by putting the toe behind the heel and pounding the heel down on the toe.

There is no other way and I don't know why she doesn't know that.

Your black socks will have fallen down around your ankles.

Find a way to pull them up over your long johns.

Never let on or in any way reveal that you are wearing long johns.

Check yourself for neatness.
Go to your desk and sit down.
Fold your hands on top of the desk.
Peek at Darlene Hudson and Casey Gahr.
Rise and sing the National Anthem.

That is how you go to school in winter in Big Lake, Minnesota.

Red Oak

If you have ever made a fire in your furnace in Big Lake, Minnesota in the winter time you already know that what you need is some nice big chunks of Red Oak.

Big chunks of Red Oak are called blocks and most of them are so big they are hard to lift , but they get better as they get older. If you live in Big Lake, Minnesota you be expected to know by the time you are seven years old that Red Oak is the best kind of wood to make hot fires in the ice and cold. But, there is a lot more to learn about how you get to where the Red Oak trees are ready to be firewood.

First, you must ask your father in a nice way if he will please take you to the woods with him so you can help him cut trees. And trim them. He will say that you could get hurt out there and that he hasn't got time to keep an eye on you and that no, you can't ride in the trailer on the way home because you could fall out or you could freeze to death. If he says, "You could freeze to death," you will know he is taking you along. If he says, "You would freeze to death," you will know you aren't going so you may as well quit whining.

Second, you will know that he is getting ready to go if it is Saturday and about 8 am in the morning when he is hitching up the four wheel trailer to the back end of the car. Your mother will be humming in the kitchen packing a lunch of about 64 sandwiches and 49 gallons of hot coffee, black.

Third, when you go out the back door to go to the toilet you will see that your neighbor, Vernon Lannon, is in the garage with a great big long saw that he is sharpening along with some double edge axes. He does not have to sharpen the wedges because you just bang them with what is called a sledge hammer in Big Lake, Minnesota. There is no way you can sharpen a wedge anyhow, because they are not supposed to be sharp. You can ask anyone about that.

Fourth is when Vernon puts all the axes and the big long saw and the wedges and sledge hammer in the bottom of the trailer. And the car chains that are hanging by the work bench. He will get in the car and wait for your father and you to come out the door. Vernon doesn't talk much, but he is known as a "goin' Jessie," whatever that means. Your father will tell you that he is also a very hard working man who will give you a full day. I don't know what that means either.

Fifth is when your mother hands you the lunch bucket and the thermos and asks if your mittens are tied on right. That means that you are going to the woods and you can go get in the back seat of the car. Remember to shut up and to stay that way.

Sixth will be getting to Eagle Lake and then turning left up the hill and through the scraggly part of the woods called scrub oaks which are really fake oak trees that never grew up very good. When you get to where there will be some chopping and sawing your Dad will stop the car and let Vernon out to show him where the stumps are under the snow so he doesn't back the trailer over them and get hung up, which I think means stuck.

Seven is looking up in the sky and learning to say things like, "We can fell this one between those two," and, "The squirrels aren't going to like this," and other stuff to make you laugh even if you have no idea what they are talking about. There will be more on this part about laughing when the work is done.

Eighth is to get the tools out of the box of the trailer and arranging them in proper order which means putting them where your Dad can find them. Then ask him if we are going to burn the brush piles today and when he says yes that means that you will be dragging the branches that are too small to be cut up with the double bit ax and piling them in the open spaces. When that happens later and the brush piles are burning you can watch the sap sizzling out the ends and that is very exciting to woodsmen for some reason.

Just before they start cutting and sawing they will tell you for the four billionth time to stay out of the way because they don't want to squish you with a falling tree. Never remind your father that he has said all that so many times before. Just nod and grunt like he does for an answer.

When it is really cold there is no sense in trying to cut trees with an ax because it is like cutting ice and the ax will just bounce off in a dangerous way. Both your Dad and Vernon know what it is like to cut ice because they used to do that for Roy Hall on Big Lake, which is Big Lake, Minnesota. They will saw up the trees that are already down and trimmed with the big saw with two ends and one of them on each end.

They get going real good and they have what is called a good eye for making eight foot lengths. If it is really, really cold they will take their sledges and wedges and split the big trunks up because they are too heavy to load and so forth. They also have things called pikes abut don't ask me what they do with them except sometimes they will have made big stacks of trunks and they will use them to help roll them into the trailer and they will say, "Look out!' quite often and very loud.

When it is lunch time they will light one of the brush piles and sit by it to dry off and we will eat a quite a few things, many of which are frozen by now. Your Dad will tell you about when he was a kid clearing land at Cable, Minnesota where he was born and raised and that he had a shack where he stayed and that during the day when he was cutting trees the fire would go out and the mince-meat pie his mother gave him would freeze and that's why he likes frozen mince-meat pies now. I never know when he is teasing me, but I think that part is true.

It is time to load the trailer and get for home when it starts getting dark and you have to be home for supper and your Saturday night bath in the wash tub in the kitchen so you better not be late. When the trailer is full Vernon will stand on your Dad's side of the car and wave his hands back and forth to show him how to get out. The wheels of the car will spin and then Vernon will say, "Rock it!" which I thought meant to go like a rocket but it doesn't. It means to rock the car back and forth by going forward and then backward to get what is called mo-something. Your Dad will tell Vernon to behave himself because you can't rock a car with six thousand tons of red oak in the trailer behind it.

Then they will both crawl upon top of the wood in the trailer and unload some of the logs. When one or the other says, "That ought to do it" they will climb down and Vernon will wave his hand some more and then they will get up and unload some more logs. When they have got just the right load left the Ford will pull out of there and Vernon will have to run alongside until they get to the top of a hill where he can get in.

When you get home they will unload the logs by taking off one side of the stakes of the trailer and pulling them off into a nice pile where they can be sawn on some more to get the right length for the furnace the next year because they have to do what is called drying for a long time. My Dad will split them up before that so they can dry quicker and not get spunky.

Just when you think you have become a good woodsman and you are standing there admiring your wood-pile your father will head for the barn to feed the rabbits and tell you it's time to go in the house for supper and your bath.

How to Spend Your Nights During Winter
in
Big Lake, Minnesota

After you have done your chores and had supper and helped do dishes by not breaking any when you dry them and put them on the cabinet top and done your homework there are many exciting things to do unless it is already time for bed.

One is to do spelling and writing which your mother will help with while she is darning socks or mixing dough if it is Tuesday or Friday.

This can be done wherever she is at because you will be at a table anyhow.

You will want to be doing this after your brother has finished his homework which doesn't take him long because he is quick in arithmetic especially and you may not be but keep your voice down anyhow as reading out loud bothers him if it is you doing it.

Another thing to do is listening to the radio all together in the living room especially Friday and Saturday nights when you can stay up later or Sunday night 'til nine when you will have heard Lux Radio Theater. Saturday mornings is when you and your mother and your canary can listen to the Hartz Mountain Canary Show where the canaries sing and your canary tries to imitate them and he gets really excited and jumps all around his cage until your mother lets him out to sit on top of the radio and he gets going really good.

Gangbusters is on Saturday nights and so is a lot of good music like Lucky Strike Hit Parade.

If you get your chores done by five during the week you can listen to the Green Hornet and Jack Armstrong the All-American Boy and if you are done by four-thirty you can hear Jimmy Allen and Speed Roberts who will send you their picture by their airplane used in their adventures.

So you can see why it is important to get home early and change clothes and do your chores and be a good worker.

There are also many books with pictures to help you with your reading and some with no pictures which doesn't help much except for practicing bigger and better words when your brothers and sister are through with them.

Some of the books will be good enough to read more than once and then there are Big Little books and cost you a dime at Warnecke's.

The catalogs from Montgomery Ward and Sears and Roebuck are the best and you can spend many hours writing down the page numbers and order numbers and colors and sizes if you know them and the best fun is the tools.

Everybody should look at tools.

And musical instruments like the coronet and trombones from somebody called Conn.

You must study the catalogs carefully and revise your lists frequently during the winter.

And leave your lists around, especially the one with a complete cowboy outfit which includes shirt, hat, belt, chaps, two guns and holster for $6.00 postage included up to age 12.

Do not ever whine even at Christmas which I will tell you about later in plenty of time for sizing things up and making good decisions.

These would be the Fall and Winter catalog.

Listening to the evening news with your mother and father at six pm is good experience because it will make you curious, especially about people.

And borrowing your uncle's National Geographics which he will gladly let you do but return them as he is what is known as a collector.

Drawing pictures after you have learned how to trace them on a window pane is also a lot of fun. There is always a chance for you to make improvements in the colors and shapes.

There will also be evening where you don't have supper but dinner instead.

This will be when you have the Lutheran minister or a teacher or two home for a meal as they are very poor and need a decent meal so they can go on with their good work.

Preachers and teachers are not allowed to get married because getting married does something to their morals, whatever those are. I think.

I don't really know about morals but being not married does not damage their appetites.

You will have more silverware and gravy than usual and two kinds of dessert.

Keep your mouth shut and your extra hand in your lap where there will be a cloth napkin.

Do not drop anything and if you are asked a question be sure to answer when you aren't chewing.

Most of the time the teachers come in pairs because that is how they live at Hadden's great house or wherever.

They will also usually be your sister's teacher or you oldest brother but try to get your folks to invite your teacher whenever you can because you will be already in love wtih Miss Hoyt and she is so beautiful and kind.

And they will talk about you and say nice things you'll like to hear.

If you hear your father down in the basement sharpening the kitchen knives on the big wheel which he will pedal to make it turn go down there and ask if you can help.

He will say yes, go ahead and pedal the wheel or stand ready to pour water on it for some reason.

Do not try to get smart and speed things up because there is only one speed that he likes.

Sometimes when you are alone in the basement you can pedal it as fast as you can go but someone will yell what are you doing down there so be careful who is around.

Also watch what your father does to things on his wood bench where there is a vice to hold things in which are in need of repair.

Or he will show you how to build a wren house or a blue bird house which are his favorite birds and mine too.

Pick your own birds, and bird house colors.

These will be made up so they are ready in the Spring.

He will also talk to you about tools and what they are for and how to use them like the draw knife.

He will also let you practice your sawing on old

boards and some hambering of nails if it is Saturday.

There is also some snow shoveling you will be able to do at night which is fun because the yard light is on and because you cannot get to the outhouse without it or a shoveled path unless you are in a hurry and don't need the path shoveled as much as you need to get there. A wide path to the outhouse is helpful because a person tends to swerve when they are in a hurry.

If you are on good terms with your brothers and sister and there is a snowstorm going on and everybody has their homework done you can play Monopoly or Sorry or Authors.

This will help you later in life.

Or you can turn on the yardlight and not go to the outhouse.

All you have to do is rub a hole in the ice on the window in the kitchen window and watch the storm and the drifting snow.

You will soon be glad to be inside and get into bed and listen to the wind as you fall asleep warm and cozy.

Uncle Doc

When we were kids we used to go to our Uncle Doc's a lot.

He lived in Monticello, three miles from home in Big Lake.

If the folks didn't happen to be taking us for dinner or a visit, we'd still get there somehow. Often times it was by walking or hitching a ride, but on our bike a lot of the time after we were old enough to earn the money to buy a second hand one. That would be one bike for the three of us boys. We developed Big Lake's first bike rationing system, but everyone knows how that works. The oldest gets the most.

Doc's wife, Aunt Lil was a great cook and made the best morning glory biscuits in the state. Doc was a veterinarian and a very good one and he was my mother's brother, which made him a Kiebel. They had four sons. Actually, they had five but Jackie died at the age of six. There was Bill, Howard, Roland or Skinny, and Dick. Skinny is the only one left now, and he lives in California someplace.

The boys are survived by some terrific sons and daughters and it seems the prevailing genes of good looks, good taste and a love of fun and good food carries on and probably will for several more generations. When we get together nowadays, it's a revival of memories in real time.

Doc was out on call a lot of the time and he had the misfortune to be in the business when that and every other business was going down hill. It was during the depression years and I doubt he ever got paid in full, but he still did all right with payment in kind. If he had had a distiller or a brewer as a customer he would have liked that. But he didn't, though it was rumored he would take the occasional drink or two. Not on the job, mind you, but pretty close to it.

Anyhow, there were four of us kids, too, only sister Lorraine didn't find the seven of us boys together all that alluring. She liked going there a whole lot less than we did and usually spent the time with her nose in Doc's National Geographic collection or obliging by playing the older folks a few requests on their very nice Baldwin piano. Lorraine was a lot more comfortable at Uncle Oscar's and Aunt Glady's farm with cousins Donna, also an accomplished musician, and David, an exact picture of Teutonic perfection. She thought he was gorgeous, even if he was a cousin.

Doc was a sentimental fool according to wife Lil. Both of the Kiebel boys were that way. People said the girls were a lot more sensible. They were right in some respects, but Aunt Laura was the exception, marrying four or five times and she never did get it right.

So, as kids we were always finding reasons and ways to get to Doc's because it was always so interesting. He had more things a kid could get into and explore and investigate and even make occasional discoveries.

For instance, there was the barn where he had his surgery. It had a genuine horse operating table in it. The thing stood straight up and the horse or cow would be led in, made to stand next to it with no suspicions whatsoever until the straps and buckles and chains were all attached and then it was cranked down to a horizontal position with the horse in it's side. Then Doc could administer the sedative and get on with the business of operating on whatever was bothering the poor creature.

You could get into all kinds of trouble attaching a cousin to that kind of table and you probably could fool a neighbor girl for a few minutes, but I don't think that kind of thing ever went on in the barn. It was just too close to the house. One of Howard's friends said you didn't need the table with some of the Monti girls, but I think that was just talk. Bill and Howard and my brother Keith probably played doctor and nurse their own way and not vet and mare or whatever.

And the medicine bottles out there in the cabinets. Doc had the most amazing collection of mystical medical bottles ever gathered together, or at least it seemed that way to a couple of seven year olds like cousin Dick and me.

We'd climb up to the shelves and look through the cabinets, take off the caps or pull the corks and take a whiff. If it smelled good, we tasted it. If it didn't, and most didn't, we'd go pewing and screwing up our faces all over the place like we had been involuntarily poisoned, for sure. It seemed as though the medicines that looked good, maybe were blue in color, tasted better than the ordinary old brown stuff. Especially the horse liniment.

Often as not, we'd find a little something Doc had stashed alongside the horse medicine and bag balm. He had a grand variety of stuff, seemingly all over the place. It is entirely possible his schemes were used in Ray Milland's movie, The Lost Weekend. In the toilet tank. In the overhead light fixtures. Some kids looked for hidden Easter eggs. We had a better game.

Which brings up another reason for going there, particularly in winter. They had indoor plumbing and we didn't. Those toilets got flushed every time we looked at them and it must have sounded like Niagara Falls downstairs when we were playing upstairs.

One of the reasons Doc spent so much time doing his inventories was because Lil didn't want any of his stinky medicines in the house. Or his stock of drinking material. She was a Methodist, I think, and a WCTU lady to boot.

I'd have to say the two families got along better than most relatives. We were always going places together, mostly to visit other relatives, but to Como Park and the Duluth Zoo and Jay Cook Park and the like. And the Botanical Gardens at Como Park was a favorite Sunday afternoon attraction. Anything that was free was on the agenda when gas was .17¢ a gallon. There were never and arguments. Not even about cars. Both the Kiebels and the Hunts were disciples of Henry Ford.

Doc had a great lawn and I now suspect it was another of his sentimental weaknesses. Lil loved the gardens, too, but not like Doc. She couldn't take the heat and the bugs. Doc thrived on that sort of environment. And manure. Hog manure didn't even bother him, and it's a good thing, being as how that was part of the trade. I don't mean he reveled in it, I just mean he more than tolerated it and enjoyed being in the animal barns or yards.

In his lawn was a rock garden. It was a real, honest to goodness rock garden with the emphasis on rocks from all over the area he served. There were flowers growing among them that would have died any other place in the world. He also had this odd shaped pool with the greatest yellow and gold goldfish or whatever. And lilypads. You could take lilypads from the lakes then and put them in your rock garden pools. I don't think that you can do that any more, but maybe you can.

He also had two or three swings and English style hedges and gardens in each corner of the yard. If you were fool enough to get involved in mowing his lawn you could get into real trouble. Those garden edges had to be perfect and there is a knack to spinning a push mower that only experience can give you as to when to quit pushing and spin the rotor on the lip. Doc was not above criticizing nine year old volunteer lawn mowing helpers. If you were to ever cut into the flower beds you could get unfrocked.

As we got older and the brothers left for the service in World War II, Dick and I had a five piece band. We also had our older brothers clothes until we left for the armed forces ourselves. This really helped our image on the bandstand. We'd sometimes practice in the living room using that great Baldwin piano.

That would usually be about the time Doc would be getting home from his calls out in the country. He had a "C" gas rationing card which allowed him unlimited gallons. When we were there making this infernal racket with, "In the Mood," "String of Pearls," Muscat Ramble," "Johnson Rag," (our theme song) or one of the other popular songs we had the sheet music for, Doc would smile, knowing he had the perfect excuse to disappear to surgery. Or, maybe he would go down to Bill Schneider's Pool Hall. Doc was a music lover with a low threshold of pain.

We all made it through the war okay, though Howie, as a paratrooper, was pretty well beat up landing in Holland. I saw Skinny on Guam and Dick on the destroyer, USS Otterstatter during the invasion of Okinawa.

Even with them all gone from this earth I still like to go by the old house in Monti. Parts of the hedges are still there and the barn is still standing. When I look down the drive it isn't hard to imagine that they are still there.

Maybe Doc's spirit is still there. Or maybe he's out on a call.

Going to to Your Uncle Oscar's

Your Uncle Oscar and Aunt Gladys have two children who are called David and Donna.

They will be your first cousins at your mother's side.

They will live on a farm on the other side of Monticello which you will learn about as you get older.

When you are old enough to visit and you are old enough to do it by yourself David and Donna will have already left there to become rich and famous. That is both good and bad because you will miss seeing them, but it also means you have got Aunt Gladys and Uncle Oscar and the whole farm to yourselves.

Aunt Gladys will want you to stay with them and so will your Uncle Oscar.

That will be okay with you as you will soon see.

There are things to do on the farm which are called chores. You can help your Uncle a lot at these if you went to stay there under those conditions and have promised not to get lost.

Some of the chores include milking the cows but by the time you get up in the morning the cows will have already been milked so the milk is still fresh and warm when you get a glass to go with your breakfast. You will drink more than is good for you.

Your Aunt Gladys will need to have the eggs gathered and will show you how.

Stick close to her as you go across the farmyard to the chicken house or the geese will hiss at you and flap their wings and the turkeys will attack you to get to your eyes.

It is okay to hold onto her hand or her skirt.

Not all the chicken eggs will be in the chickenhouse.

Some chickens like to hide their eggs in odd places and some cannot help it if they lay their eggs outside as they must do it whenever they can.

Brown chickens lay brown eggs.

White chickens lay white eggs.

Some people say brown eggs are best because they are what they call rare.

When baby chickens are new they are called chicks and you can scoop up a bunch of them as they are soft and noisy and always running around hungry.

And pooping.

A setting hen can hatch as many eggs as she is big enough to set on.

A brooder is what you want if you have a lot of chickens but it is not as good as a setting hen because a brooder doesn't worry about how you are doing if you are a chick. How would you like it if you were raised under a brooder?

When your uncle feeds the pigs it is called slopping. While he is doing that it will be your job to yell sooey, sooey, sooey real loud at the pigs so they know it is time to eat and won't miss their main meal. All that yelling may not be needed at all because the pigs always know when it is time to eat, but it makes your uncle laugh to hear you yelling your head off, so do it anyhow.

You will also be able to put down hay from the hay mow by not falling through any holes up there. You will learn to like to say mow and you can use it several times a day.

He has got a special stool with only one leg that he sits on when he leans against the cow to milk and he sings pretty songs to them. If the cow doesn't like what he is singing she could do some fancy kicking and that upsets your uncle. I think he is more upset that the cow doesn't like his songs than he is about the kicking.

The cats always come to watch him and listen to him singing and he will shoot a bunch of squirts to them so they can go lap, lap, lap as the milk hits them. He says it is quite comical to see them do that and when you aren't there to help him do the milking he still has their company.

When he lets you in the milk room it is to turn the handle on the cream separator which must be done at a special speed or you will wreck everything, so pay attention to his instructions every time.

Be sure to stick close to your uncle on the way back from the barn to the house. A goose never forgets and will also try to grab you on your bottom if they can get close enough. Your uncle will tell you about his father's stories about making the geese in the old country eat a lot so their livers get extra big and then there will be lots of liver worst. That is his story anyhow, but I don't know how a goose's liver gets worst.

The cream from the separator goes in a cream can

which is called a shot gun can and everything gets strained for flies.

Then those are poured into milk cans with big handles. That would be the cream that is poured in, not the flies. The next morning the milk cans are taken to the Land O'Lakes Creamery in Monticello where they give you money for the milk and cream and any eggs your Aunt Gladys will have sent along with you.

There has to be dogs to do chores right when you go to the pasture to get the cows. Cows eat out in the pasture all summer long but have to be brought home to be milked unless they come by themselves and smart cows will know that and the others will just follow them along to see where they are going.

If they don't come home by 4:20 PM you will want to go to get them with Shep, a Collie, or Blackie, a black dog.

The dogs are happy to do this and to get the chance to nip at the fat cows.

Don't let the dogs make the cows run or it will wreck the milk they are holding to be milked.

I don't know how you keep dogs from making cows run.

My uncle does. He carries a switch which he whacks his overalls with if the dogs start to make the cows run. They will stop quick and lay down on their bellies with their paws out in front and look at you for further orders.

Mostly, he can make the dogs stop from running the cows by just saying Shep or Blackie and that's it.

If you are alone and trying to get from the barn to the house you can take a horse to the watering trough under the windmill.

Turkeys will not go for your eyes if you are on horseback.

When you get to the watering trough you just slip off the horse on the other side away from the turkeys and climb over the trough into the back yard. There you will be safe. Don't worry about what to do with the spare horse. When the horse get through drinking he will wander back to the barn and stand there until he gets let in.

If it is Monday you can help get water from the cistern which is a huge hole in the ground under the house and is made from big rocks.

The water is rain water which comes off the roof into the drains and down into the cistern. Rainwater is what is called soft and is very good for washing hair as well as clothes and other items needing special washing.

You will not have to help weed the garden because if you wanted to weed a garden you could have stayed home.

It is not all work at your Uncle Oscar and Aunt Glady's farm.

Like their strawberries and cream and homemade honey from the bee's houses.

On cornbread which is still hot and has some homemade butter on it with a glass of ice cold buttermilk and fresh peas, too.

And fried chicken which could have been laying

your eggs yesterday.

With chicken gravy.

Your Aunt Gladys will tell you to never fry chicken in anything but a black cast iron skillet which looks like a frying pan.

And homemade lard.

And six hundred green onions.

And German wilted lettuce salad which is just that because it gets wilted when you pour hot grease on it with sugar and vinegar and nothing else will work.

And your German grandfather who can't speak much American but likes you anyhow.

He lives in the room off the parlor and stays there most of the time or on his own porch where you can sit together and look at each other and smile a lot.

He has his own kitchen which you will want to get to know as he will send you to get things from there for him and for you, too.

But, he will want to get his own cheese that he makes himself from his homemade cottage cheese and keeps in the cupboard wrapped in wax paper and some newspapers around that.

It is green because it is moldy and is just ripe or right, I don't know which.

When he has got that peeled and has washed his own big purple grapes he will let you sit on his lap in the big leather chair and look out the window at the lake below and he will feed you cheese and grapes.

And some tastes of his sweet homemade wine.

It is made from the same kind of grapes left over from last year.

He is quite old but his beard is soft when you squeeze him.

He will show you picture through a thing that you hold up to your eyes and the pictures are on a stick at the other end and they look like they aren't pictures but real.

They will be of his home in Germany before he had to run away to Wisconsin in a little sailboat.

And his mother and father and brothers who never came to America and he never went home.

But he doesn't cry.

Not the way we do anyhow.

You will do this with him as often as you want.

In the winter time if you are there he will sit in the parlor by the great big stove with windows in it and do what is called snoozing in German.

Do not bother him then or try to show him anything you are doing because you will not have his attention.

In the summertime you will have to go down to the lake and find things to do by yourself.

It is okay to go swimming in Bertram Lake if someone older is with you or if you can find a place where they can't spot you.

Be careful of Bertram Lake because it has a DROP OFF and the turtles will get you.

If you walk with Shep he will lead you down a path along a creek made by cows between Bertram Lake and Long Lake.

Cows never do their duty right on the path but don't step off to one side barefoot without looking.

At the end of the path you will spot a pool which has lilies in it made from Lilly pads.

There is a dam there which holds water in the pool unless it rains too much and only a stream goes over the top.

That is called First Creek because another one from Long Lake past Mud Lake is called Second Creek.

There are many beautiful flowers all along First Creek and you must follow the creek or you will never find Long Lake at the end. My big brother said he thinks the Indians planted all the flowers there and then we took them away from them.

First Creek runs over rocks and boulders and bubbles and the baby turtles sit on the rocks in the sun and take their naps in the afternoon.

Shep will wake them up for you.

These are called mud turtles and not SNAPPING TURTLES which will grab your toes and never let go even when you take your red jackknife and cut their heads off. I have never had to do that.

SNAPPING TURTLES are as big as a washtub and they will quietly lie on the bottom of the lake and wait for you to walk by. You cannot see them ahead of time and there is no warning because they're down deep, just waiting and waiting and waiting for you. I have never been attack, but many others have been.

In the Spring at the dam on First Creek you will be able to see huge fish called carp who are having babies.

Some people do not care for carp but they will not

bother you no matter how big they get.

Not like SNAPPING TURTLES.

If your uncle is not too tired from his plowing or planting or cultivating or harvesting or doing the chores he might row you out on the lake in one of his boats which he rents out to people who want to go fishing. He has boats on both lakes and fishing either one is called an experience.

What you do is the same as you do when fishing in Big Lake, Minnesota only you do it from a boat.

What you want to do is take some worms from the garden, or grasshoppers from the meadow or some grub worms from the sawdust pile near the machine shop or the wood pile.

You will do drop-line.

That is when you fish over the side of the boat with your black line and lead sinker along with a hook on the end holding your bait.

When a fish eats he calls it nibbling.

When they have tasted enough and decide they like it they will grab hold of the bait and make a run for it so no other fish will try to get it and they can go eat in peace.

You are old enough to learn the difference be-tween nibbling and real biting.

If you pull up your line too soon while a fish is nibbling you will end up doing what is called losing a fish. Those fish are always the biggest.

If you haul it up while the fish is biting you will do what is called landing a fish.

The difference between the two is what you catch.

If you go fishing and stay out until nearly dark or are fishing too deep you will catch bullheads which swallow the bait but are very good eating if you can get someone to skin them.

To skin a bullhead you must lay him belly down on a stout board and pound a nail through his head.

Then take your red jack knife and cut the skin around the head and then down his back to his tail. Now, you must take your pliers and pull the skin all the way off.

Ask someone else to do the rest. If no one will do that, then you must give the fish to the pigs as they like them very much. Be sure to take it off the nail and the board first.

You will also want to learn to do trolling for Northern Pike.

This is done with a cane pole or a casting rod with a great big minnow on a spinner or spoon. A minnow is an orphan fish.

You will learn the language soon.

Whenever you get a big fish on you must yell I got one and reel or pull in the line.

The first few times it will be weeds or a dead branch but you are still learning so it is okay.

When you do get one you will know it.

Your uncle has caught many, many of them but he still loves to do it and gets excited even if you are the one catching a Northern which can take a while and you must keep him from doing what is called running under the boat or you will have a mess and lose him.

Uncle Oscar says there are Northerns in those

lakes bigger than you are.

Another way to go fishing is on Second Creek which I told you before runs between Long Lake and past Mud Lake.

It is hard to get to because you must walk about 400 miles over hummocks which is very difficult to do and talk to your brother or cousin Richard and carry your pole and bait at the same time.

Second Creek has got trees and willows all along it and it is hard to find a clear spot to use your cane pole.

But when you do, Oh Boy!

You will get your limit of sunfish, punkin seeds and bluegills every time.

Put them on your stringer or if you forgot it or don't own one, put them on a willow branch.

I will show you how to prepare a willow branch to hold them so they don't get away when you put them back in the water to stay fresh 'til you take them home to Aunt Gladys to fry for supper.

With fried potatoes.

And fresh biscuits with jam or jelly from the cellar on them or more honey.

You will be very hungry from all that walking and talking and hauling in fish all day.

Farm dogs can eat fish bones but town dogs can't.

I don't know why that is.

By this time you will know a whole lot of stuff including why farmers and you must wear straw hats.

If you don't have a big straw hat you will die of

sunstroke on the farm in a matter of minutes.

And burning green grass keeps mosquitoes away but only if you stand in the smoke which is not too great as you will cough your head off.

When your folks come to get you be sure to hug your grandpa and tell him you will come to see him soon if your Uncle Oscar and Aunt Gladys want you to come back.

They will.

Pickles, Sauerkraut and Other Stuff

If your are in school you will miss out on many nice things, but you can also catch up on Saturdays in the fall.

But when your mother will do her pickles is before you go to school anyhow so you will get to wreck your back permanently picking small and larger cucumbers by the bushel basket full or as much as you can carry.

Before you start it will be a good idea to take your wash tub and scrub brush up to the pump and pump some water in it. Don't bother to rinse out the tub.

Pick the cucumbers before it gets too hot, but no the big ones or your mother will just throw them out and look at you really stern. There are plenty of them for all kinds of pickles at the same time.

Scrub the cucumbers with your scrub brush and put them in the basket. Now you must empty the tub and rinse it real good. Pump fresh water in it and rinse the cucumbers and put them back in the basket and take them to the house.

Open the cellar door and take them down to your mother who will inspect them. When you get enough cucumbers down the cellar your mother will send you back to the garden to get the grape leaves you will need from the grape vines.

These must be the smaller ones which are high up on the arbor so you will need a ladder from the side of the barn.

Put those grapes in a grape basket for some reason and take them to the pump and wash them one by one on both sides and let them dry a little on the grass.

When your mother has got things ready she wil
bring down the vinegar and alum.

And dill weeds.

Vinegar is very good so taste that first.

If you taste the alum first you won't be able to
taste anything else for a week, but taste it anyhow
when you have had enough vinegar to last you for the
day.

If you have got a bottle of vinegar you will have
some on hand for when you want to shoot corks. If
you don't even know how to do that you get a bottle
that fits the cork you have and put a teaspoon or more
of baking soda in the bottle and then pour a little
vinegar on top of that and then put the cork in and
point it at the cat.

Shake it quick.

It will shoot real good.

I thought everybody knew that.

Your mother will have scalded the quart jars in the
copper boiler on the kerosene stove in the back room
because she doesn't want to heat up the only cool
place in the house, the basement.

You can pack some grape leaves in the jars and
some garlic in each one too.

Your mother will have to do the rest about the
vinegar and alum, but maybe she will let you push
the cucumbers into the jars and jam them down
before they become pickles.

But you have to get some dill tops and that is
really what makes dill pickles, so that goes into each

jar at the bottom and then some more at the top, too.

I don't know what goes on after the jars get the rubber rings put on and then the tops tightened so ask you mother to do that part.

Then you are done with dill pickles and she will take care of all the other too, such as dinner pickles and ice box pickles and later on sweet pickles with cinnamon and watermelon rind pickles.

You will be eating a lot of green beans which can be eaten from many ways of fixing such as you'd expect so no two days are alike or you'd get sick of them.

When the tomatoes are ripe you will be able to pick plenty of them for canning down the basement, too.

You will not eat many of them while you are helping your mother because there are so many and they have to be scalded to get the skin off and that is not all fun as you will find out and makes your fingers sore from doing so much, but your mother will tell you they will taste good next winter.

So try to think about that even if it won't work.

You should always obey your mother.

It is too bad you don't know about canning peas but that is very hard to do as is corn and there is too much work to those for what you get out of it.

Whatever a lug is what your father will get several of with peaches and pears in them and you will eat as many of these as you possibly can before they get made into sauces, which are canned along with everything else in sight.

During canning season you will be so stuffed you will be happy with potato salad which is made with secret dressing that goes like this:

Make some fresh mayonnaise with egg yolks and fresh lemon juice and vinegar, pepper, mustard, salt and boiling water but not too much.

Put in some extra mustard and some sugar and fresh dill weed.

That is the secret of potato salad dressing.

There is a big stone crock which will hold all the sauerkraut you can make from cabbages in your garden and more.

You must wash the crock and the board cover which is already white from sauerkraut and fits inside the crock as does the rock.

Do not try to lift the rock as it is very heavy and has to be washed too for what reason I have no idea as it never does touch anything which it would get dirty which it wouldn't anyhow because it has been washed for the last five thousand years at least.

The cabbage cutting board is an invention which will slice your cabbage just the right thickness.

It is put on top of the stone crock and braced against the wall. You will need to hold it because you will be pushing cabbages back and forth across the blades until you run out of cabbages or the crock gets full.

Your Dad will get to do this, but he will show you how to do it if you act old enough.

Beware!

Do not get your fingers anywhere close to the

blades or you will lose them all slice by slice and there will be blood all over the sauerkraut and everyone will be mad and scared.

Only a sissy would wear gloves no matter how clean.

Every once in a while throw in some coarse salt.

It is called rock salt for some reason, but it is not the rock you are salting, it is the kraut.

There are many other things that go on, too, but I forget what they are.

Root Beer

When you have been to first grade and get out of it you are old enough to be in charge of root beer.

You should have learned enough during that time to do it just right.

First of all you need a big wash tub.

Take it out to the back yard but near the back door because you will need lots of water to wash the bottles and that is the closest pump by the kitchen sink.

Bring all the big bottles up from the basement and the smaller ones too.

Wash them up really good inside and out.

You should have enough bottles for sixteen gallons of root beer.

Turn the bottles upside down to get any surplus water out.

When you have done that haul the bottles back down cellar and line them up so you can fill them.

Put down lots of newspapers because this can get messy.

Get down two boxes of bright new bottle caps from the shelf overhead as you come downstairs from the pantry.

Try your bottle capper by pushing the handle down and then let it up.

It will work.

After your mother has mixed the Hires Root Beer Extract and sugar and yeast with the water in her big stone crock you will use a hose to syphon the root beer from the crock into the bottles.

You will have to have some kind of clip on the

hose to shut it off if you are a good guesser about when the bottle is filled just right one inch from the top.

You start the syphon by sucking on the hose.

You will get a snoot full of root beer that hasn't bubbled yet but is real sweet anyhow so that's okay.

You could get a little more by saying oops I lost the suction and doing it again.

You must put the end in the bottle when it is running and then shut if off and go on to the next bottle and so on.

Your brother should hand you the bottles.

Keep doing this until you have all the bottles full or you run out of root beer from the crock.

The last ones will probably have more yeast and they can blow off the caps in the middle of the night and your Dad will wake up and say I thought I heard a shot.

Anyhow, now is when you will see the mess.

But get all the bottles capped.

Do the big ones first and then adjust the capper down to the shorter bottles.

Regular beer bottles work real good too and they are usually dark colored and the root beer keeps better I think.

Put the root beer in a dark place to get the yeast working.

Clean up the mess and tell your mother when you're through if she isn't already standing there shaking her head.

You must wait five days before it is ready.

But it can be used after three days in an emergency.

An emergency can also be an occasion, whatever that is.

Sometimes the yeast will just die a natural death and never fizz at all.

Keep some in the ice box at all times.

When you do run out it will be fall and time to go to school again.

How to Watch Your Mother and Father

Glossary of Terms:
"Need it", is Adeline Hunt's secret process for preparing bread dough.

If you like to learn and are a good worker you should keep an eye on your mother and father.

They have many good habits and are interesting to watch and practice copying on a daily basis.

There is no special place you have to be to do watching either your mother or you father.

That is because they are everywhere you look.

Night or day.

It is much easier to watch you mother at first because she is usually in charge of your upbringing and you will be spending a lot of time with her.

Start early because you are never too young to learn.

Everything your mother does ends up in the kitchen.

So start there.

Unless it is summer and too hot to be in the kitchen.

Baking is the best to watch because it is always something to eat.

You will know when it is time for baking because the pans come out and have to be greased except on Saturday when butter is used.

Then there is a big basin or tub to mix the flour and yeast which is mixed with some secret ingredients by hand.

When it gets stiff your mother will quit stirring

and attack the dough with her bare hands to need it.

I don't understand why she says she needs it but when she gets done she puts it on top of the warming oven to get the puff up.

She will go on to do other things so there is no sitting around.

Rolling out pie crust is another thing to watch because it is rich and more dry and is hard to manage without wax paper to work on and to use to turn it over into the pie pan.

Then there is always whatever there is that is going inside the pie such as apples, cherries, apricots or peaches or even rhubarb.

But I've never seen a pear pie.

Then the top crust which you must pinch the edges together to hold in the juices and turn to trim with a knife and cut flower pictures on the top.

Your mother can probably hold the pie from underneath the pan and trim all the edges with one turn.

Watch carefully as that part is tricky.

If it is a cake, no matter what flavor there will be eggs and sugar and a lot of stirring and twirling and mixing to be done before it goes in the pan after which they are put into a hot oven and then start the frosting.

Then there is a lot more egg whites and sugar and I don't know the rest but it all get delicious from the egg beater you will soon be licking if you keep your trap shut and just look into your mother's eyes when she is done and she says Oh all right you can lick the spoon.

This also works with ice cream.

Now she will be ready to punch the bread dough and put biscuits and loaves into pans and shove in the oven.

The fresh bread is very hard to wait for but the raw dough is something your mother will always offer you so learn to like it and thank her for it since it is better than nothing.

And sometimes there will be enough cookies to treat an army as she says but knows they last exactly one week unless your father's sweet tooth is acting up and he must pack it with many cookies of different kinds.

That is where the fresh milk comes in and your father can pack many cookies if there is enough milk, as you will see.

Your mother says it is no wonder I have to bake ten days a week but she is smiling as she scolds and wipes the flour off her hands on her apron which started out life as a flour sack.

If you are clever and don't get smart she will let you cut the cookies with what is known as a cookie-cutter before Christmas or Thanksgiving or a jar top any other time.

It is a good idea to ask your mother if you can try a little bit of cookie dough because that way she will say yes and it is much better than snitching some.

A snitch once in a while is okay but don't think you're getting away with it because you aren't.

Learn to accept that.

Cooking is different than baking but not as much fun because there is very little sugar in meat and potatoes and vegetables.

There is much to be learned in the cooking business, too, so watch how everything must be clean such as food and pots and pans even if you fry it to death.

Washing dishes is okay to watch but you are never too young to dry them as you will soon find out.

One problem is that most people get into dishwashing and wiping while they are still too young to realize what has happened.

But that is okay because there will always be dishes to be done.

Sewing, patching and darning will be in the dining room or living room.

Some of it looks like fun like crotcheting and making pictures on dish towels with colored thread.

If there are yard goods on top of the sewing basket on top of the sewing machine in the dining room and it is not near Christmas or a birthday of yours you will find out that somebody in your family is due for a new shirt or a nice blouse for your sister.

Or another pair of short pants.

If it is Monday you will find your mother in the basement.

You will know for sure it is Monday if your basement air is full of steam and it smells like soap and your mother is pumping on the handle of the washing machine or stirring clothes in the copper boiler or wringing clothes from one tub or the other as they get rinsed.

And then she will take the cellar steps two at a time to get the clothes line which are put up around the oak trees to hang everybody's clean clothes and bed clothes and towels on.

You be sure to help her when she gets a load of clothes in her basket ready to hang up by staying out of the way or handing her clothes pins to do all those with.

If your mother is back in the kitchen ironing clothes it is Tuesday.

You can help her by learning how to sprinkle clothes.

You do this by putting your fingers in water and shaking them up and down over the clothes and then roll them up tight and put them back in the clothes basket.

She will add more sprinkles to the clothes as she irons them and puts starch water on your father's shirts.

Your sister should be watching too, but she will be busy or in school.

Your mother will only stop ironing long enough to make you a sandwich and get your own milk out of the icebox unless it is winter and then get it off the cellar stairs.

She will have an apple or pear and some cold coffee.

If she is ahead on the ironing she will make a pot of tea.

Recipe for hot tea:

Put some green tea in a tea pot.
Pour boiling water in 'til it is full.
Cover the tea pot with a towel for a while.
(About five minutes, I think)

Try to sit still for three minutes while you drink some.

Shuffle your feet under the table and say ahh.

Wednesdays are when there is a pile of rugs on the cellar door which end up on the clothes line.

Your mother will whack the daylights out of them with a carpet beater but you can do this or they will wait 'til your father gets home to do it.

If you own a Bissell Carpet Cleaner you are lucky because they are very good to use and your mother will go tearing all over the house behind it looking for carpets to clean upstairs and down.

Sweeping and mopping are things you should know by now but if your mother ties a dish towel over her hair and wraps up her broom in a rag you will soon see her chasing cob webs which are really spider webs.

I have never seen a cob so don't ask me what a cob web is. It is a spider web no matter what anybody else says.

She says that if we got rid of that moulding there would be no spiders or webs on the ceiling.

I don't know what she is talking about half the time.

You will notice that whatever you mother is working at she will hum or maybe once in a while whistle a few notes.

That is because she is always happy to be doing something for you or your family.

It is a hard thing to learn how to do and mean it, but it is something to strive for as your mother will tell you.

Like shooing the neighbor's dogs away without hurting their feelings.

II Watching Your Father

Watching your father is also done in many places at different times.

The best time is when he is whistling with his teeth.

He will seldom jump on you then although he may want to do some Dutch Rub to anyone on hand wrestling because that is his favorite sport and he is known to give a friendly tussel when he feels like doing it.

Mostly he will save his famous wrestling holds for your Uncle Doc who thinks he can whip him, but can't because he always says he has his glasses on or you'd better watch out.

Your father will tell you about champions name Bronco Nagurski and Vern Gagne and his favorite because of his blind speed and remarkable footwork Cliff Gustafson, The Norwegian wrestler he saw in St. Cloud.

Or a boxer named King Tut.

Listen carefully or you will get the same story over.

His name was also well known.

When you have got all the information you need and if he is busy at building things in the garage or down cellar you will notice what good care he takes of the tools he uses.

There is a saw, a hamber, a rasp, a draw knife and some screwdrivers.

And a hand drill.

You will often see your father looking at the mail order catalog to see about some new tools.

He will say to your mother that he sees where the price of wood and metal lathes is at an all time low and she will say really that is nice and he'll never order anything because there are more important things in life.

Whatever happens you will be sure that your father will be able to make do.

That is the same thing as doing without only it lasts longer.

Most of the time your father will be making his own version of what he'd like to buy.

He has got an electric motor to do this and it will do whatever he says it will.

Believe your father at all times.

In the summer it is better to watch you father because he is highly successful in making things and doing daring projects with very little to work with.

In the fall he will let you go with him on Saturday afternoons to where he cuts wood on shares.

That means for all the red oak he cuts down and trims and saws up by hand for his trailer to haul home and he has to make piles of wood for the owner to come to get in his truck.

Then when the wood gets home he will make a big stack of it and split it all up to dry out for a year.

This means you had better stay busy or you won't have enough red oak for the furnace.

When you get old enough to haul wood you will know all this and how to load your wagon or sled because you have been watching and helping your father.

Watching your father chop down trees and use his big saw will make you very proud of him and he will let you help him burn brush when he is ready to do that.

It is important that you get his tea at this time because he will want to drink some tea while he is watching so the fire doesn't get away.

He will also teach you all about squirrel nests and mushrooms and when ducks are getting ready to leave and the noise they make and the pheasants call or the partridge beating on his chest and chipmunk's storehouses and muskrat houses and beaver dams and fox dens.

And skunk holes.

These are all good things to talk about in school since you didn't go anywhere in the summer.

Neither did anybody else but the teacher always asks anyhow and hopes somebody did because she didn't either.

In the garden when he is hoeing you should watch in case he tells you to take over.

When he says take over that is the signal that he has to go feed the rabbits and ducks and pigs and also means that you have learned hoeing well enough and already have spent a lot of time watching with the feeding chores and he trusts you.

So do your best work always.

This is no time to whack a cornstalk instead of a weed or start day dreaming or acting up.

Do your job and do it well.

In the winter if you are still real young, both your parents will put you in the box sled and haul you down to the pavilion where the high school boys and girls basketball teams will be playing or in the KOTM hall.

I don't know how much it costs to get in but it is a good thing that kids are free or you'd never get to see any games even if it is Monticello that Big Lake is playing.

Don't even think about running out on the floor at half time.

The kids that do that have no manners at all and their parents probably drink beer or worse.

Everybody knows that but if it wasn't for basketball games they wouldn't get the chance to discuss it or shake their heads and talk behind their hands to their friends who are nodding and yesing behind their hands right back at them.

Bill Snyder and Lyle Smith will make two or three baskets and Big Lake will beat the enemy 16-12 or worse.

Bill Snyder gets the center jump everytime because he has got springs in his legs.

Mr. Ostroot the Superintendent will be there and everyone respects him because he is fair and Big Lake always wins.

Anyhow, there are always times when you don't want to be too close to your father especially in the winter time when he doesn't get home until supper because the roads aren't plowed and he has to make his own tracks to deliver the mail.

He will be very tired and just wants to eat warmed-over supper and fall asleep in his chair and do some snoring before he gets to bed.

Don't pester your father because he could bark at you even if he doesn't want to.

Another time not to watch your father is when the car won't start because it is too cold and Dick Rosengren has to come with his wrecker and pull him to start or if that doesn't work he will have to put the car in Dick's warm garage to thaw out while your father sorts the mail.

It doesn't matter what the weather is like because the mail train will always get through and Conner Bradfield or Mr. Putnam will have been over to the station to get the mail bags for Big Lake and all the farms around and there are plenty.

Your father will have to go out no matter what.

Another time not to watch your father is when he gets stuck in the driveway and has to shovel himself out.

You could take a chance if you want to try to help

by going down the basement and getting a bucket of wood ashes and take out to him but you are bound to spill some and your mother will say things like where are you going with those ashes or look what you're doing to the kitchen floor or stop tracking that stuff through the house.

Once you get it outside it will blow in your eyes or your father won't think it will do any good because there is still too much snow under the frame and he is what they call hung up.

A Ford is too powerful for its own good in snow.

But he will take the ashes anyhow and send you back inside to warm up.

I have heard some very loud cussing which is really why he will want you to go back inside.

Sometimes your older brothers will go out to try to help or your mother but there still is no reason why you should watch everybody get mad at each other.

A stuck car is very confusing.

If your father has got to lay down on the snow and try to get his chains put on to get out from being stuck watch from the window upstairs.

Do not go out.

It is not safe.

Watching your father whittle wood after chores and supper in the winter is very rewarding but don't expect to learn how to do everything he can do.

Not even when you're old enough to have your own jack knife.

A red one.

He will try to show you how to whittle a wooden

chain from a piece of flat pine wood.

Or he can make a wooden ball in a block of wood.

Or a dancing man on a stick.

Or a monkey with all his arms and legs flopping just like the dancing man.

He will grin at you and say how do you like that when he has got his project done.

Then he will take the projects and paint them and give them to poor children on the route for Christmas.

You won't know them because they go to country school.

PART TWO

**Part Two consists of
Short Stories
Poetry
and Other Stuff
Written in Big Lake,
Alaska
and Ireland**

POETRY
OF
ALL SORTS

Although it is frequently
called
"Free Verse,"
it isn't.
You've already
paid
for
it.

The following poem was first published in the May, 1934 issue of the Minnesota Journal of Education.

That's right. The May, 1934 issue.

In the Evening

In the evening when the cows are milked,
I like to watch the sun go down into the hills
And hear the robins sing their last songs.
Evening is the best of all
When the moon is shining on the lake beyond
 the hills.
It looks just like strings of gold upon some
silvery glass.
That is why I like the evening light.

Rodney King Hunt
Third Grade
Big Lake, Minnesota
Lillian Crofoot, Teacher

In The Craw of a Crow

In the craw of the crow
One will find
All the roadkills
Of mankind

Eschatological material of Babylonia,
On tablets recalling Gilgamesh
from the temple library
of the god NEBÛ

And from the palace
of the Assyrian king
Ashurbanipal of Nineveh in the
Assyrian empire

And this crow knows
the goings on
of all the Sumerian kings
in the land of Uruk

With the deluge
Enlil had ordered
and the answer to Enki's question
"Why hast thou commanded.....?

She followed the other deluge
from the height of her Zorastoral tree
and hopped to the Christian branch
to see the latest killings

From both black eyes
she could look upon
this people described as dust
for the nature of the covenant people

And listened to Maimonides,
Moses ben Mimon
strengthen Yemenite Jewry
to withstand persecution

Later saw them scattered
bloodied, disemboweled,
spat upon and jeered
for the good of the world through galuth

Auschwitz, Buchanwald
Triblinka, Mengel
Hitler, Himmler
and lesser Teuton heroes

Watched Alexander do
what he did best;
pouring blood on rivers
and the Med

Craft not an army
without a God
Said kings of England and Italy and Spain
Crusade ye the Holy Land

Killed the heathens,
Killed the infidels;
Drank their blood
from pots and pans

She was privy to the
forces of Chaos,
watched St. Anthony shout down
the temples of Karnak and across
the Nile, Luxor

Shate upon the horses
of Napoleon and the
polemic penises
destroyed in Egypt

Winged high to view
the pyramids of Mexico,
An obsidian knife cut out the heart
of an innocent fool

Welcomed then Cortes and Pizzaro
envied their bright gold a bit
and ships with loot,
pitied the ruin of Machu Picchu

Annishanabe to the north
led west from the east to pine
 and die of dead buffalo
on church fired prairies

Flanders fields,
Somme and Stalingrad
were picked clean
along with her comrades of other species

Or did she fancy Gallipoli,
or sweet Crimea
with a taste
of fine Sepoy bowels?

Industrial strength child labor,
Territorial imperatives
with a manufactured touch of
profits from powerless pains

Crumbs to Somalia,
Rwanda: spell thy name.
Chernoble and
the Pacific bombs of France.

Rising from resin,
smoked in glass.
Worthless parents,
hopeless children

Sits on the last limb
On this earth
Watching for the
final termite

In the craw of a crow....

THE DEATH CYCLE

I'm sorry about
the death of my brother.
He thought I understood.
I didn't.

UP FROM DOWN

When your day just won't start
Stuff your nose with sauerkraut,
fill your ears with beans
then
do things with your jeans.

Put your finger in the air,
Smile and turn and jerk
Two-step like a Texan
then
Slide and glide and stomp.

Eat a pound of coriander
Drink a pint of stout
Microwave the Barbie Kevin
then
Shake your booty all about

Tickle your sister with
a thistle
When she's sick
and you'll find,
all your troubles down the toilet
then
you're feeling fine.

Dignity

There is something about
the fastened cuff of a
long-sleeved white shirt

covering a young man's arm.

What it is I don't know
but there it is
in its complete statement
trying to be understood.

Could it be the wrist or hand
emerging straightforward
from the cacoon of puberty

born to a lost discipline?

Or is it just the aching
to see once more
in my lifetime

Dignity?

INVITATION TO A
FORMER LOVER

At Kylmore Abbey near Lenane
Awaits a haunt for thee
Her erie vapours dripping blood
Seething and raging convulsively.

Dark recesses within the soul,
Foul of mind, of hair, of breast,
Cancered loins of loveliness
Await you dear, in death.

In your scorn shown yesterday
A caustic fire was lit
Dying love bred life to her,
And murder to you yet.

Old Shadows

The toxic shadows
cast by dead landlords
around our town

Have infected some of
the people in
this water-soaked place

Their particular malady
is sometimes characterized by the
uncontrollable tugging of the forelock

And awkward, but not forgotten
bows and curtsies
in certain company

Say no more,
Say no more

Hear,
O Israel!

From Proverbs 6:16-19 I call you to stand before
the abomination of your acts

From your new land has come
a leadership which taunts My words:

> Haughty eyes, lying tongues,
> Hands that shed innocent blood,
> Hearts devising wicked plots,
> Feet that are swift to run to mischief,
> False Witnesses who utter lies,
> and the sowing of discord among thy
> brethren.

And Hillel said, "Whatever is hateful to you,
do not do
to your neighbors for that is
The entire Torah as I have written;
I am the Lord who created your soul,
and the soul of your neighbor,
and he is you.

As you encounter me,
So must you hear me.
And just as your passion loves the past,
the word for now is no different

than the word then,
or there would be no word for all:
Hear O Israel,
the Lord our God,
The Lord is One:

Yom Kipper effects atonement
for your transgressions against Me;
Yom Kipper cannot atone
for your transgressions against your neighbor
or your fellow man
if you have not appeased him.

Singer of My Psalms
Victims of the holocaust,
Oppressed, now oppressor,
Save thyself, redeem thyself!

Redeem thyself!

Ireland

Beneath this earth,
Beneath our rock,
Beneath the waters,
Both fast and still.

Slow thunder rolls,
And anger flashes,
Welling up,
Then subsiding.

Seventy four miles,
From Kinsale town,
Northeast by East,
At 74 degrees.

In the roundtower's
Morning shadow,
It hides
Within the glen.

Beneath this earth,
Beneath our rock,
Beneath the waters,
Both fast and still.

Two Hundred
Thirty
feet
straight
down
is
my
heart.

Shortwave at Night

There is absolutely nothing quite so enchanting and mind boggling to me as having a radio capable of receiving short wave transmissions. I sit in the darkness of my study and as I turn the dial the world's greatest show begins.

One need not be a linguist to understand the pagentry of nations being presented by hundreds of broadcasts on hundreds of stations as they pass by, stopping like a colorful merry-go-round at your fancy; each band with its own style and purpose. The mysticism of Middle Eastern music with its point-counterpoint. The Israeli Symphony; the Opera from La Scala or as it is being presented in Vienna, Japanese Jazz, Indian Rock, Irish Western and German martial or Oompapa.

Radio Marti to Cuba and Fidel coming back with his own, all in the rich and delicious rolling of r's and slashing s's. BBC comes from dozens of on-the- spot sources providing the best in unbiased news reading, the best in sports reports from Commonwealth countries and uncommon countries. Rhythmic, clipped Oxfordians from the Upper Sudan and Kenya, from Mozambique and with the hope for death of aparthid in South Africa, a not so gentle British reminder of the predominance of their power stemming from the land of poets, 'this England', Christian missions appealing for help for the starving in Somalia.

One is always snapped back to the reality of the Latin nations of South America by the remarkable music of that continent, San Paulo, Rio de Janerio. Raciffe, Bogata, Quito with the instant visions of Mancho Piccha, the dark width and dangers of the Amazon, rain forests spring to life, slash and burn. A Festival breaks out anywhere at anytime and all the time.

The winter music and comment of Siberia, the new cities, the new settlements, the new lands, the new countries, all striving to be heard to be understood, old champagne in crystal glass, too soon drunk, too soon broken. And, the realization that democracy is not a hot-house flower, that it cannot be misunderstood and then force-grown under glass any more than it can be won in the streets with speeches. Musing, dear God, that they can see it as well as they feel it, but soon I'm swept away by the Volga, the pictures of Stalingrad in ruin, the beauty of the Hermitage in Petrograd, and the cheapness of a good youth going for the American way in Moscow.

Each time there is the realization of truth in that these are people from stations around the world who tell their stories, play the music and act like themselves instead of being portrayed as something they aren't by our own wickedly-intentioned media who make greed a virtue. The refreshment as they can and do tell their own stories and play their own music their way, without commercial interruptions, distortions and while it may very well be propaganda, it is not a propaganda of selfishness and salesmanship.

It is ethnic, it is purposeful, it is real and it is almost spontaneous since you control their entry and exit. It is the best of each station's best, it is on show for the rest of the world. It is sincere and honest, refreshing, inspiring, sensual and exotic if you'll just close your eyes. It's there twenty-four hours each day, it's in color and it's a dream coming true, it's free of all fat.

One can always hope with short-wave radio at night. One can always enjoy it without cable and it goes everywhere with you. You don't need a schedule or even knowledge of who is on when from where. All you need is the magic black box and the ability to twirl the dial to happiness.

Local Hero

The Levade

Local Hero

In the village of Big Lake, Minnesota in the year of our Lord 1936, there lived a man named Everett Johnson. His true age is unknown, but copies of a long-faded baptismal record found in a shoe box in the rear of Archie Larson's store in Orrock have been carbon-dated, lazar-zapped and methodically mutilated by scholars from the University of Minnesota Agricultural College.

This thorough system of guessing places Everett's age at about 11, making him a known and practicing Libra. This discovery alone indicates poor parental planning and created the paradoxical dilemma of a dreamer born to the farm. While statistical evidence such as this is not uncommon, it was aggravating to his father and mother.

Though Everett's birthdate is still more conjecture than absolute fact, his diary was also recovered, wrapped in a yellow oilskin Southeaster beneath a monumental manure pile on the dairy farm which was his last known residence in Sherburne County.

Bearing in mind that Everett, the story teller, was, by his own account, a sporadically public school educated man, he was well read and possessed an uncanny left-brain-right-brain symbiosis; the dreamer and the actualizer.

The diary tells that his childhood was no more or less happy than that of his peers. He played infrequently and worked at his father's side in the fields and barnyard from dawn to dusk. Neighbors were distant and only slightly acquainted; the one room school near Melvin Enger's (Public School District 30), being seldom attended was not the social factor that it might have been. School records in the Superintendent's office of District 727 confirms his love for reading, none-the-less.

As was true of most Norwegian farmers of the era and area, field work was largely accomplished by teams of horses and Johnsons had four such good animals. Everett's father increasingly left the care and feeding to him and thus the obvious bonding of boy and Equas caballus. He minded the chores. Witness:

"Seen to the hay from the mow for Nell, Nancy, Mildred and Harold. All in good shape and pulling well enough. Harold acting frisky again, but it is Spring. Nell very interested, but Mildred tries to hock him. Ma in family way."

Everett's growing up with and around those horses led him to riding them, sans saddle, from age five forward. His skills were seldom if ever seen outside the farmland of his home. But, that skill is recorded by two keen observers and respected horse-men, George and Ed Everett, who wrote in the best sellers, Horses in Big Lake, "The Johnson boy could probably mount and ride any horse he wanted, but stuck with his own and rode like the very Dickens. He seems a part of the horse and he stayed without punishing horse or himself."

As drought and low corn prices posed a parallel but juxtapositioned economic paralysis, Everett's keen interest in horses continued undiminished. The driving force was revealed in a brief entry, smudgingly dated August 3, 1937.

"Must get quarter dollar together for Sherburne County Fair in Elk River. Show bill says they got the trick riders from Buhl. Must learn their ways and methods to improve own efforts. Must ask Mr. Swanson can I clean his barn Sunday to get ready cash."

And, Everett got there, leaving us this message of his dedication to learning horsemanship.

"Buhl must be a wonderful place. Got good study on their Belgians at Fair. Good jumpers and just dainty. I never see such a thing. Will try jumping Harold over wagon tongue tomorrow! Must fix hay baler first."

There are many references to his frustration with training efforts, the lack of time, lameness of self and horses, but grim determination was acted out in efforts to read even more of the Haute Ecole of Equitation.

It was the sudden tragedy of the fall of 1937 which could have, should have, destroyed his dreams. He writes:

"Pa has sent the horses to auction. There's no money from field crops like as corn, wheat, rye, or oats. No use to plow and harrow and drag and plant nothing. Milk no good either, but cows to stay. Pa says horses was getting broke down from too much jumping and galloping in the north 40. Howard Olson has new collie pup."

About that same time, Clarence, Everett's father,

got a postcard from Arvid, his cousin in Minneapolis. Written Sept 12, 1938, it said: "Clarence you best get yourself down here with me and work on construction with Bolland Brothers as they are building Northern States Power Co. steam plant. Or WPA."

Then Everett wrote the note.

"It is already been two days since Ma and Pa and the girls has went to the cities. Think Pa glad to leave me to care for farm and cows. Lonesome, somewhat. Wish I had a horse to talk to. Phyllis Anderson dropped off a nice cooked chicken. She acted kind of funny in the barn. Need kerosene."

Things abruptly changed for the better in Everett's honest blue eyes. It was October, 1938.

"Have had a vision, as Ma would say. Walked to town today and spied a book at August Peterson's on Norwegian Riding School. Mr. Peterson let me have it for promise to clean up alley and stack root beer bottle crates, dig out basement, put out ashes and put up screen windows on house."

The book, no longer in print, was a catalyst for Everett's efforts. His excitement was intense as he described the contents. Other books, too, were referred to in glowing prose.

"Finished part on Hunting Deer on Your Herford and it bonkles the mind.

Peri Hippikes, Xenophon and Ordine Di'cavalcare by Griso good if you got horses, but I ain't! Give me Ole Swanson's Daring Bovinity (Ole Swanson was the inventor of the cattle curb bit.)

Everett would train his Holsteins to the rein and

English saddle, to leap, to assume noble positions, to be all that they could be!

For a raw farm boy from Big Lake, Minnesota, this was heady fare! No more would the 2 x 4 be carried into the barn to gain his cow's attention.

From Everett's notes, his devotion to Ole was complete, but not limited to Ole's teachings alone. From him he began to believe that obedience is not best bought by subjection or force. According to him, the cow should be taught this obedience by rewards, tid-bits, praise and petting, but to be cautious about loving promised of anything else. Whips and spurs, therefore, were to be used in moderation and then only in broad daylight. The animal must be taught to think, but riding a Holstein is a refined art and a process that brings about the closest possible union between man and his cow.

Everett was soon to discover that Swanson's eldest brother, Enoch, had written intense and knowledge-able detailed descriptions of bridling and the proper use of saffles, pillars and inovative curbs for Angus cattle. Enoch had produced the much quoted Novel Ways of Training Gurnseys to the Western Saddle. It was a brilliantly executed work, but the ideas were reactionary rather than progressive since Enoch advocated artificial and forcible training methods, i.e., the ice tong theory of keeping the animal's head up and the controversial paired axe handle girth stays, so popular with Prussians.

Despite the success of these authors and others with impressive documentation, Everett stayed with the Holsteins. His decision was based upon their perfect and nobly shaped body, graceful movements, the desire of the animals to learn, vivaciouosness, courage, hardiness and perseverance. The expressive head, a high set neck (for a cow) comparatively low withers, a rather long and developed back ending in muscular croupe, clean limbs with strong, articulated joints, beautifully shaped hooves and well veined udders. At 155 to 165 cm tall, the Holstein walks with a spring and has unusually high knee action.

Everett came upon the works of an earlier Herford fancier, Freihere Von Oeynhaussen, who had outlined a training program, which he, Everett, was certain could be adapted to the even more sturdy Holstein. Those program notes became the creed:

1. The high art of riding Herfords (Holsteins) must never be understood as Haute Ecole alone, but compromises all three types of riding which are:

a.) Riding a cow, without upsetting her natural balance, at uncollected paces along straight lines, called "straight forward riding."

b.) Putting the collected cow through all paces, turns and movements, while maintaining perfect balance; called "Campagne Riding."

c.) Putting the cow that is kept reared on her hind legs with her haunches bent deeply, through all of the usual paces and jumps, as well as the unusual and artificial ones. Carried out methodically with utmost exactitude, agility and skill, and developed to the highest perfection, this type of riding is called "Haute Ecole."

If Everett was anything, it would appear he was methodical. He knew the first year was to be spent getting the cows to trust him and to develop their suppleness and dexterity. This meant riding forward to the bit, to cause the cow to strengthen the back legs, making sure they were kept close under the body. This proved difficult until he developed the "Johnson Athletic Udder Supporter" which worked well, but unfortunately, reduced milk production. Then, and only then, thus trussed, could Everett assist the cow(s) in working through turns, figures, and volte; first at a walk, then at the trot and canter.

Cantering exercises had to be developed systematically. Everett found they could not be started too early or continued for too long, lest the overtired Holstein lose her innate pleasure in leaping forward. Trotting, trotting, and more trotting became the essential of the advanced rigorous training schedule. That rhythm had to be maintained as the utmost preciseness in riding one track figures, doing corners, doing tours advanced to two-track figures. The "shoulder-in" movement as introduced by Gueriniere was stressed to improve the gait and using half-passes, "croupe-in" exercises and "renvers" to help the Holstein to become even more obedient and more supple.

The second year also began the short-rein work to further tuck the back legs underneath the body. Riderless, the cow can better lower the hindquarters so that much of of her weight rests on them. This, Everett noted, allowed a higher knee action of the forelegs. It also became apparent that it was good preparation for the movement known as "Piaffe" and an indicator of which cows were best suited for the exercise of the "Cows Above the Ground" near phenomenon.

The Piaffe, is, of course, the basic Holstein stance which can be captured by watching the normal Holstein at rest, head tucked to chin, right front leg raised, left rear leg matching in pose.

From this develops the "Pesade" where, on the short rein, the cow stands smartly on her hind legs, front legs folded nicely and in a "reared" position.

Next came the "Courbette", groundworking on the short rein, where the gallant Holsein leaps into the air with forelegs and hind legs gracefully poised beneath the body.

The "Levade" was performed in the same manner as the Pesade, but with Everett in the saddle.

The "Capriole" is the Stag Leap of the cow jumping world and a strenuous maneuver in which the cow must be constantly aware of the consequences if they allow their legs to splay while coming down from the apogee of their effort, and, with Everett in his modified bibbers, it was, from all later accounts, a magnificent sight.

After three years of near isolation Everett was surprised by a lost stranger who chanced upon the near overgrown driveway. Still making diary notes, Everett relates"

"A stranger from Duluth(?) come by for directions to Zimmerman as I was working on Shimar of Becker. Mr. Costello ask could he watch and I told him he could, but mind his oxfords. He said he could make me a movie star like Tom Mix. He was crazy about what me and Shimar of Becker was doing out there in the yard. So I give him a show of short rein and the saddlework with Princess Kay, then Bella Donna of Elk River and Queen Mary. Mr. Costello said we should be in the State Fair, but I ain't going down there where sin is so plentiful. He says we'll do it here, build a great big arena and all that. Phyllis Anderson come by later with a cooked chicken. She is still acting funny in the barn."

The stranger was intent on carrying out plans to make a star of Everett and money was no object. The West Sherburne Tribune accounts the following brief.

Cows to Get Palace

The Tribune has learned from neighbors that the Everett Johnson farm is the site of a magnificent arena for some trained cow riding.

Mr. Robert Swenson, our Orrock cub reporter, has visited the enterprise and writes the following:

"The artistic taste of Msrs. Johnson and Costello is manifested in this beautiful building designed for them by William "Bill" Putnam and William "Bill" Blackhurst.

The main front (there are three fronts) faces a small pond or swamp. From a high and massive base it rises in four stories that are divided by double columns and ends in a round corner structure. In front of the cupola, crowning the steep roof, is a globe topped by an eagle, and flanked by allegorical figures.

The stone balustrade supports vases and trophies from the winter bowling league. The facade is very beautiful, with its harmonious proportions and decorative cow sculptures. Behind it hides that jewel of architecture, the manege of this unique riding school, which fills the whole building. It is 250 feet long, 52 feet wide and 52 feet high, decorated entirely in white and bathed in light. Forty six columns support the gallery. The forefront of the hall is dominated by the mayor's box which is surrounded by a tympanium with sculptured rural scenes and a statue of Mr. Costello.

Editor's note: Young Mr. Swenson has been known to exaggerate. Thank you. G.M.

The Sherburne County Historical Society notes of the grand opening sums up a good deal of the details so painstakingly cared for by the volunteers of the Lutheran Ladies Aide Society of the Eidskoog Church. It reads as follows:

"The Grand Opening performance was planned as a masked ball. However, the blindfolds ordered from Johnson and Beech Store failed to arrive on time.

8,000 wax candles on burlap covered raised milk pails lit the hall. Coffee and ice cream spoons engraved with the Seal of Sherburne County were given to all ticket holders. Curiosity about the festivity had reached such a fevered pitch that tickets were counterfeited and sold at a high price. Unscrupulous ticket collectors had resold the tickets just collected to sightseers. In no time the spoons had all been given out.

The Mayor's box was occupied by the honorables of Big Lake, Orrock, Santiago, Nowthen, Crown, Soderberg and out of courtesy, Monticello.

The manege was tastefully decorated with cornstalks, gourds, potato vines and patchwork quilts.

A ladies carousel was planned to begin the occasion, but bad luck struck Inez and Bertha Edberg as their rig slid into the ditch in front of the Nels Pierson farm on the Elk River. Their ordinance was damaged.

Music by the Svea Hill Ramblers played what seemed to be variations on themes by Sebelius, The Peer Gynt Suite and another local favorite, "Whoona, Whoona, Whoona" by Neal Person. The Becker Polka Devils played the entire "Belsazar" by Handel in bolero time.

Events included a quadrille of the herolds, gentlemen driving carriages pulled by six brightly painted Jerseys, the gun races with 10 and 12 guage shotguns on baby buggy wheels drawn by matched heifers."

Your Historian,
Emma Davidson

Everett's great coming out was also Everett's going in. It was the first and the last spectacular. The bright lights and big press notices were not to his personal liking. Mr. Costello left for Duluth, somewhat unstable it seemed, as was demonstrated years later by his opening of a frozen Chinese food business. Everett turned his attention to classical cowmanship with its characteristic nobility and splendor. On October 9, 1940 he wrote:

"That there makes a nice barn and the cows sure

- 261 -

like it a lot. Must put gutters in the menege. Hay up to balustrades. Phyliss Anderson come by with a cooked chicken. She wanted to see allegorical figures. She is very smart and now I know what it means. She don't act funny no more."

In today's gaudy and ostentatious display of life style one might wonder at it all and ask, "Why?"

With little or no blaring publicity remaining, one must search for an answer.

For Everett it was breeding and art. Man and Holstein were fused into a singular artistic personality that developed according to its own laws and displayed the beauty regardless of a certain lack of public appreciation.

A vision of beauty was made real by a visionary, Everett. It was imparted to the onlookers as lightly and as effortlessly as a leaf floating down in the fall breeze.

It was, as Kenneth Nelson said so simply, "A joy!"

EPILOGUE

And so, one might ask, what ever happened to Everett Johnson? The story is veiled in psuedonyms and was only recently cleared under the Freedom the Information Act.

In 1943 Everett enlisted in the United States Army, took basic training at Fort Riley, Kansas where, when it was revealed that he was a celebrity cow-rider, the Post Commander commissioned him at the rank of

Captain and assigned him to G-2: Army Intelligence.

Captain Johnson was immediately sent to North Africa where he was dropped by parachute behind enemy lines to observe the Italian Camel Corp. He wrote on November 13, 1943:

"This place is called Somewhere in Africa and so far there ain't nothing in Somewhere. Camels all gone to war. Seen twelve or so kneeling in the desert. Arab man said they was praying. While they was praying somebody snuck up on them and tied gas cans over their pretty humps. What a dumb thing to do."

Later, in 1944, Everett was reassigned to the Red Ball Express along with Manny Shapiro of Monticello. Together they amassed a huge fortune in surplus goods sold from Army trucks to the French and German population as they followed General George S. Patton and the 3rd Armoured Division into Berlin.

Everett and Manny, Manny and Everett! The team stayed together and took their discharge overseas and began their adventure in show business. Manny was wild with an exciting concept.

July 29, 1945:

"Manny wild about show biz. He don't know it ain't all it's cracked up to be. He don't know the heartache and tears.

Babbette come by with cooked chicken. Manny acting funny around the villa."

Eventually, Everett was called to Hollywood, U.S.A. where he became Bobby Breen's manager and an advisor to Walt Disney Studios.

Though he had returned to Big Lake only a few

times for public appearances in the High School Gymnasium, it is thought that he, as perhaps a master of disguises, has been with us on many other instances, though undetected. There were other sightings.

Said our California connection, Doris Jennings, "It is to my personal knowledge the Everett Johnson, or E.J. in the trade, is in LaPunte, Arizonia where he resides with his companions, June Allison, Doris Day, Donna Reed and Jim Nabors."

What's in the future for Everett, Big Lake's brightest star? No one knows, but you can't get a taker if you bet E.J. and Big Lake won't get together again.

Adam's Apple
A Gift for a Teacher

This whole thing needs an explanation. As far as I can recall, my involvement came about very quietly and, as if I expected it. I assure you, I didn't.

I'm a teacher. Of philosophy, and the area is logic. You know the drill; the human mind manifests itself by thinking; thinking depends on words and language; language is a false communicator used to obtain mastery or promote subservience; the loss of freedom and the delegation of choice rather than the expression of it. Truth, mien herr, uber alles.

So I don't expect much, and other than lecturing on what I believe in I don't do a lot of unnecessary communicating. To some this makes me appear a mystic or at least a deep thinker. I'm neither.

One thing I do believe in is river fishing, but nothing fancy. There's enough sophistication in everything else. I like to walk along the bank and try to read the deeper holes for fish in the eddies with a bobber and a minnow.

So I was doing this kind of thing three days ago on the Elk River in Sherburne County near Big Lake, Minnesota. I was about four p.m. and relatively cool, and would you believe, no mosquitoes. It was near the corner across from John Person's old farm and near Donna Larson Eddy's place. I could hear the tractor I'd seen with the hay wagon behind it picking up bales from the field. Some birds, a dog fretting about something, but otherwise, very quiet with an occasional water gurgle among the rocks in the river.

It's a very narrow backwater where I was, sort of a

cove, but uncommonly deep, perhaps 20 feet. I'm not into geology so I don't know why, when the rest of the river going by is only three to five foot deep. I have had good luck there, so don't ask where it really is. What I just gave you is mostly smoke.

I'm not an excitable person, but I should have reacted differently now that I remember what happened. I was sitting, or more correctly, straddling this big maple log and this guy walks right across that 20 foot deep water of the cove!

He walked right up to me from the woods on the peninsula and across the water and says, "Good Day." Nobody says good day around here that I know of, but it seemed all right under those conditions.

And, I thought those conditions even seemed all right.

So, I said the same thing and he asked if he could sit, he said "set" on my log. I asked him to please do so, which he immediately did.

This person was wearing a Co-op cap and ordinary stripped bib overalls. Osgosh by Gosh, they said on their label. He had eight colored marker pens in his top pocket, arranged in the regular spectrum and in the correct order. A person in logic notices those things, and that his black high top dress shoes were neither wet from his trip, nor muddy, nor scuffed from twigs, branches, rocks or razor grass. His shirt was white linen of excellent quality.

He then told me in a well modulated voice that he had noticed quite a number of large pike swimming at about three feet below the surface where I fished and where he had walked to get to me.

It didn't surprise me that he told me that.

My students, those keen intellectuals out East, would have said I was cool.

The gentleman told me his name wasn't what he was known as because it been updated to Adam. He admitted, without questioning, that he was fond of that name.

His linen shirt had no collar.

I told him my name and where I was from, which was right there in Big Lake, but was vacationing from Boston. He asked where that was, or rather if it was anywhere near Zimmerman, which he had heard of. I said it was and didn't feel badly at all, because at that moment, I thought for sure Boston was near Zimmerman.

And, friends, this is where the talking became surreal.

"And, where are you are you from, Adam, if I am from here, are you, too, or also, as the syntax may be?" You can tell I was really was quite irrational and not making much sense.

"That would be correct for now," he said, but added he didn't actually feel like it.

"Is this your home, then, as we say?"

"I, Adam, can be safe at home beneath that earth; beyond the trees yonder, over and out."

"I see, it is underground you have your home or cave, as preferred by many now and years ago?"

"Exactly. It is, every bit as you have so eloquently and accurately described it. A cave indeed it is not as such so much as one would or could provide the word ' burrow,' to its description." He then calmly asked, "Would you mind turning on your marvelous wheel attached to the stick with the odd looking string, which is nearly invisible, and seems to be attached to a large pike on the other end?"

"Not at all, Adam!" and I did, landing the fish at Adam's high top black dress shoes, which broiled the pike thoroughly in a smoke filled moment.

As I bent over to the fish, Adam was headed back across the river, turning once en route to wave good-bye and tell me that, "I will enjoy your company again the morrow."

"And, where will you be?" I asked.

"Where we meet, new fellow, where we meet," he answered, and disappeared into the ash-wooded land.

In my room at the motel that night I tried to recall how Adam looked and talked and whether or not that afternoon had really happened. In reflection, I found nothing extraordinary in either, though I felt I should have.

I left early the next morning to drive around the lakes, both Mitchell and Big Lake. It was very satisfy-ing and very early, about five-thirty and the fog and mist were over the reeded shallow section and on the swamp at the fish trap. I was near the strip between the lakes known as the point, a favorite spot of mine for duck shooting, and, of course, making out in high school.

I mused that that a lot of water had passed

through the channel since last I had been there. It was at that point that I saw Adam. He was at the channel, standing on the railroad ties that held the sand back.

From that distance of about sixty yards I could see his feet were moving to an unheard syncopated beat. In his hand he had a cane which he swished to a fro from one side to the other and behind his back, twirling it like a baton. In place of yesterday's co-op cap (Midland Feeds, I remember now) was today's straw boater. The rest of his ensemble remained the same. Had it been dark and had I been unable to see him I could have homed in on the smell of hot creosote from the railroad ties being sizzled by his dancing high top dress shoes.

His routine stopped with a pirouette and a buck and wing finale with the boater ending up on the hook of his cane, which in actuality was nothing more or less than a niblick of unquestionable vintage.

My applause was subdued but sincere and I offered the congratulations due him.

"Adam! Good morning! What a surprising pleasure and entertainment to boot, so to speak, meaning of course you furious unaccompanied assault upon the ties of Burlington Northern!"

"Thank you! A mere nothing. What have you obtained in your view is but a common right to sense the sun for the day."

"What brings you here, Adam?" I asked.

"A habit, more or less, I suspect. I have found myself here each morning for quite some time, though it is unclear to me how I could answer your next obvious question."

"And that has to do with your rather remarkable dancing."

"As I thought," said Adam. "What we have here is not to be confused with anything I could possibly explain. Let us just say that my admittedly remarkable hoofing is as surprising to me as it is to you. I Just don't know what overcomes me. It happens here on these railroad ties and occasionally when I traipse between here and there on the Soo Line. Yes, I will break into a veritable outburst of near triple paradiddle rhythms of unspeakable origins and frenzy which quite exhausts me."

"I'm not at all surprised and appreciate your candid confession. Shall we stroll while your fine shoes cool a bit?"

"Ah, yes. I confess I have no exact place to stroll since my time here is taken up by myself alone."

"Oh," I said. "I'm sorry Adam. I didn't presume to interrupt."

"My meditation?"

"Well, yes, I suppose that was what I might have thought had I been able to think of it."

"Fear not. No offense taken, young man."

"I should have known. There is a certain religious aura about you, not to mention your ability to cook fish and go boatless across water of no mean depth."

"That might be true and I think you would like to believe it, whatever it was you meant by religious."

I decided it was best to skip through my mistake of mentioning religion and my curiosity about his mysticism.

"So, Adam, are you involved in something or is it hypnosis and altered states?"

"I, as one, am nothing more than an example of many. My friends and I are secure in our knowledge which presently tells me that I should not be standing in this water in my good shoes."

"Why, for instance, are you standing in the water instead of atop it as was yesterday's mode?"

"I haven't any good reason, I suppose. I simply forgot to step on it instead of into it! Nevertheless, I should react to your previous question by saying that my religion and our medicine are secular."

"I'm sorry, Adam, but I think you said your religion was secular?"

"Oh, indeed, that is ever so true, without any question my religion is but a toy I carried forward from childhood and you would do well to admit that such faithfulness allows for the aforementioned security."

"What would I call this if it has a name and is a faith?"

"I believe you could call it naturalism and animism with a wakan-like power of the supernatural which accounts for all things in the universe on an impersonal basis."

"You said, 'our'?"

"Ah, yes, I am in the habit of saying "our" when there is more than I or me. Am I wrong to say 'our' or even 'we'?"

"Oh, no. I simply did not know you had a family

or were, perhaps, with friends or companions."

"There are many of us near here, though we are seldom seen."

"Or, at least, admitted by others than myself to having been seen!" I said.

"Your perceptions are precise and our being invisible is not physiologically a standard."

"And, that means?"

"It means, young fellow and friend, that we practice not the awaiting of revelations, but rather we promote them through our potent rites."

"So, most people around here have chosen to believe they didn't see what they did see?"

"A magic without equal, an occult without induction, an agreement with nature and without black connotations!"

We walked along the sandy beach, I perplexed with my hands behind my back, Adam taking swipes at the puff tops of ripening dandelion blossoms on the verge with the brass-headed niblick.

"Where are we going?" I asked, thinking he was leading.

"Where are we now?" he asked.

"If you don't know where you are, how did you get here?"

"I know where I am, David. I am with you."

"That makes sense," I said reflectively.

"And what is that which makes sense, as you say, and I now find myself saying, or rather asking, since it appears in my mind as a question?"

"That makes sense which you said about knowing

where thou art for thou art with me, Adam."

"That is clear."

"Are you going to tell me who you are or where you came from, or both; since I have already confessed as much of myself to you?" I asked.

"Of a certainty, but first we must get to Huntington's pasture, though it is not a pasture now, but nearly filled with dwellings of most unusual formations. Do you know where this place is?"

"I do, for once I lived near the famous pasture and favorite place of all in my early and later childhood. But, why, Adam, do we go there?"

Adam looked at me as though he had just discovered I was not at all bright.

"That is where I live!"

"In one of those houses?"

"No, but let us go. I'll meet you there."

"Don't you want to ride with me?"

"No, David. I'll be there when you get there."

"Yes, I'm sure you will, but where? It is six square blocks in size, scope and sequence!"

"I'll be looking for you when you arrive," he said and was gone.

I managed my way to my car and turned on Highway 25 toward Monticello and then southward on the graceful, curving streets of the new addition that had once been Huntington's pasture.

And, there was Adam, standing in a jump suit in one of the few hollows not yet filled in and sold and built upon. There, too, was a gnarled, massive willow stump. In my recollections, fast flashing, I could recall such a mammoth willow tree standing beside a cherished, lovely pond.

"You arrived quickly!" I said.

"A relative term with small and larger meaning."

"So it is," I said, "as are small and larger. I am absolutely amazed that the willow stump remains behind you!"

"Should it be in front of me?" he asked.

"No, I only meant that it surprises me that the stump remains. That it has survived decay and destruction."

"It shouldn't, friend David. We seldom eat willow. Much too bitter and contains flourescein, C H D phthalic anhydrule with trace FO flourate; 20 12 5 2. Virtually nutritionless and terribly fattening for anything so desperately sour. Some use it for a purge, however, with only a low degree of success, or so I'm told.

"You don't eat willow?"

"No, do you find that strange?"

"No, not at the moment, Adam."

"Do you like my new clothing, David?" He asked, innocently enough.

"I was going to comment."

"It is all the rage, is it not?"

"I can hardly be a judge, Adam. Leisure suits were the last thing I saw around here."

"But, David, did you not know leisure suits cause cancer of the brain?"

"Perhaps that is why the fashion faded." I said.

"You were going to say, 'died out', but took mercy

upon me, I know."

"That is true," I admitted.

"With all these truths you have discovered, would you now like to learn more in answer to your recent unanswered questions?" he asked.

"Of course, Adam. I asked who you were and where you came from."

"And, David, with a strong but polite hint at what I was planning to do here. You have waited patiently. Come, then, with me." he said.

"And you were going to say, 'walk this way' and took mercy upon me, too."

"Another truth, David. We are absolutely endowed with them!"

Adam made a sweeping gesture to indicate I was to go forward and as I took one step an oaken door in the earth appeared and opened onto a short chute. Adam plopped himself upon it and I followed, entering a glowing chamber as the door above us closed solidly.

The room was tastefully done in roots and subterranean ferns, the floor of natural rock and humus. There were seats from the Crossroads Mall and River Park in Monticello as well.

"Very nice!" I said.

"You were amazed and surprised that we did not eat your willow stump, but you are less than impressed to be four meters deep in Huntington's pasture!"

"I'm beginning to think the only things that surprise me are the things that survive we humans!"

"I agree. Humanity has become a tragedy. Modern man invades the world of the wakan man."

"And, if this is so, I have just invaded the wakan world and am doomed to death?"

"Oh, gracious, mercy, heavenly days! I should say not! You are our guest!"

"'Our guest?'" I asked. "I see only you here."

"That is so, but let me introduce my wife and friends."

"Am I to speak? I still cannot see them?"

"If you feel their presence or if they speak to you, which they will assuredly do. They're anxious to meet you."

"How many are present, Adam?"

"As many as I will introduce, David. And with that, and first, my wife, the former Lady Sennui."

"I am delighted to know you, Lady Sennui."

"Former Lady Sennui," she said. "I am now Atum's wife, David."

"Atum?" I said, questioning the pronunciation.

"Yes, Atum. Atum the All as in Memphite Theology."

"Of course. I should have known," I replied to the husky voice of Sennui.

"This is En-khafet-kai, and his wife, the former Lady Sesheshet."

"I'm glad to know you." I responded.

"Call me Ernie," said En-khefet-kai."

"And this is our little friend Seneb the Dwarf," introduced Lady Seshesehet.

"I'm glad to meet you."

"Well, then, for now, that finishes the introductions," said Adam, or Atum.

"If you will excuse us, David, we shall leave you and Atum to yourselves."

"I'm honored to have met all of you." I replied, sounding less bewildered than I really was, I hoped.

Adam said, "Now that they are gone, let me tell you that you have made a very good impression."

"Thank you, Adam. Are all of you really ancient Egyptians?"

"No, we aren't all ancient Egyptians. Some are, some are latter day."

"Are you at liberty to tell me the who, what, where, when, why and how of things as they appear to me at this very moment in your vault?"

"It is not so much 'at liberty' as 'of necessity', dear David. You are the only person we can tell. We are certain of that."

"Why me?"

"Is your Social Security number 473-10-0004?

"It is."

"So, we chose you."

"That's it?"

"That, as you said, is it."

"Let's start with that. A Social Security number is how you meet people. No internet magic, nothing but a Social Security number?

"We don't believe in computers, as I shall later show you."

'Then how was my number selected?"

"Have you, in your studies, read or heard of

Macat and the Kings?"

"Vaguely recalled from Egyptian Mythology."

"Well, we'll let that go. It isn't all mythology. However, Macat stands for truth, for order and for justice. His was the supreme governing will and right dealing in all things. Not a set of rules, but interpretations based upon the situation. So you aren't a random number.

"Right. I remember now."

"Then let us sit upon one of these splendid iron and concrete seats with the terrazo effect, so popular in your land.

As we sat he offered me a Diet Pepsi and took another from the wall behind us for himself.

"I find many things to my liking here in the land of Big Lake."

"I do too, It's a wonderful place to have been raised. How did you happen to find it?" I asked, relaxing.

"We didn't really find it, David. We more or less were thrust upon it."

"Could you tell me about that?"

"I will not go into detail for it would simply take too long. By the way, do you favor Greeks?"

"Do you mean, do I like Greeks?"

"Yes. Do you like them, the Greeks?"

"Yes, I do. Very much."

"Good, David. We have several of them here, too. Fellow named Demetrius from Syria, Mausolus from Halicarnassus, lovely girl named Theano and then there's Praxiteles who designed these quarters for us.

not his particular forte, but good enough for us. Not exactly Mnesiclesean effort, but good enough for us, as I just said."

"And, you all came here. To Big Lake, by accident, if I heard you right."

"Not exactly that, either. My memory is not as good as some, David, and correct me if I'm wrong, but is there a temple near here called Elk River?"

"It is not at all a temple, but there are good citizens who live there, I'm told, though I've not met them."

"I see. Was there a temple then called the REA? Or a god, Rea?"

"There was, and is, a power plant there which supplied electricity to a cooperative known as REA."

"And did the REA have an atomic powered electrical energy plant which developed a crack in its reactor?"

"Yes, Adam, your knowledge is accurate. About 40 years ago."

"Would you hold onto your Diet Pepsi with one hand and your section of this marvelous concrete bench with the other?"

"Of course. I'd be glad to."

"And, perhaps you could close your eyes."

At that instant, with eyes firmly shut, there was a steady increase in the glow and sound not unlike a dynamo accelerating, but with no sensation of movement."

"You may open your eyes, David."

As I did I saw we were in a huge, rock-walled cavern and I could hear and then see, water, radiant, angry, colored water running down the rock walls.

"My God! Where are we?"

"I am the last person you should ask about where we are. My old friend, Mursilis over there calls it Monticello. But then, you know what he did to Babylon, so I'm not so sure I believe everything he says. There's an old saying about Hittites bearing gifts."

"Why, then, Monticello?" I asked.

"Is there not an atomic power generator there?"

"Yes, there certainly is," I replied as the mental dawn began to break.

"Is there not a rumor of storage leaks and denial by the God of Power, know far and wide as "Northern States?"

"Well, yes, sort of like that, though they now call themselves 'Excel' now," I replied.

"And, does not the pretty, colored and pulsing water look different?"

"Yes, it surely does."

"We are below the power plant, David."

"This is bed rock, Adam?"

"This is bed rock."

"Oh, my God!"

"And Big Lake is the corner of the triangle and the safe spot of the low pyramid. Draw the lines in your mind, David. Even the lost Atum can do that!"

"And, this is why you are here?"

"No, David, this is how we are here, not why."

"The radiation from Elk River and Monticello?"

"And Three Mile Island and Chernoble. That is

another triangle, a massive triangle, a monumental pyramid to the dead."

"I cannot believe this!"

"We aren't asking you to believe it."

"But, these plants were shut down!"

"I know that, David, but not before the neutrons got us up."

"Then it doesn't matter. It's too late!"

"Not really, David. Not really. There is something of a crisis. Let us return to Big Lake. I feel safer there. Just do what you did before."

"Care to come with us, Mursilis?" he called.

"No thank you, Atum. I rather like it here," came the answer.

The sound of the dynamo wound down and we were safely back in what I started to realize was their bunker.

"It's not a bunker," said Adam.

"Then what is it? And, kindly warn me when you have read my mind."

"It is our place to be. No offense intended, David"

"While you were out, you had a call," said the voice of Sennui.

"Really, it's getting so busy! Who was it, dear?"

"It was a ka call."

"Oh, dear. Is it time, then?"

"Yes, Atum, it is time." said Sennui, in a calm voice.

"What, pray tell, is a 'ka call' if I may be so forceful as to ask and I do because I have this sinking feeling that something is going to happen which will profoundly shorten my life span! Something, perhaps, like my beginning to glow like a neon sign! After our visit to Monticello, I mean it!"

"You are in no danger. We put a little something in your Diet Pepsi that neutralized the radiation effect."

"I want to thank you!"

"Ka, David, is for the lack of a better term in your world, soul, which functions after death. It is an expression of the total aspect of God's being. So when the ka call comes, which is rather pretty to say, don't you think, we must assemble our various peoples and collectively give our analysis of the situation, our recommendations and our farewell. David, good fellow, after that, in brief, ' we are out of here,' if I may paraphrase an expression I heard at the Broken Spoke on my first day in Big Lake."

"All present, Atum," said Sennui.

"Very good. As you must know, we will be leaving beautiful, safe and friendly Big Lake in a matter of a few minutes. You have all had ample opportunity to gather your data. Now, make your calculations while I give David our sociological summary."

"David, this world will not last past a second set nuclear errors. I've told you the belief we have taken from wakan and modern man's preoccupation with self destruction."

"You have, Adam."

"Then, too, you know about the insanity of the children of war, the politicians and their equally insane military minions. Not unlike our man,

Mursilis, though we learned through his assassination."

"I do know, Adam."

"There are armed weapons in space, there are missiles on the ground, in the ground and under the seas. And, there are dedicated persons whose sole job it is to kill, by action and retribution, every living creature on this planet. They have the authority to do that! I believe their stupidity is only superceded by their arrogance. A 'defensive shield,' indeed!"

"Yes, I know that, too."

"David! You cannot let them do that!"

"I, David, cannot let them do that? And, just how, in God's name, am I, David, going to stop it?"

"In God's name, thou shalt, David, go forth and slay Goliath!"

"You're mad! You're absolutely crazy!"

"No, David, not mad, or insane, or even angry. Just tired of modern man."

The deep voice of Mursilis, just returned from Monticello, spoke: "David, we shall tell you how and then you just do it. It is the least you can do."

The voices of the others joined in, agreeing with Mursilis.

"Then, let us explain how it works so David can get on with it and his life as teacher of Philosophy in Boston, which is near Zimmerman," said Adam in an Atum-like voice. "First you," he said, pointing toward nothing I could see.

"You will use maser-microwave projections of coherent gamma rays," said one.

"They will transport bayron breeders of hyperon multiplets which penetrate the molecular epiderm of silicon chips and magnetic tapes." said another.

"This will destroy the memory and the communication systems within the operatives of the electromagnetic spectrum." instructed a third.

"The projection and transportation of the coherent gamma-rays is governed by the electrovalency of the photons in discrete subatomic particles," said the first voice.

"These are converted to electroluminescence which will produce secondary radiation," said the second voice.

"Then, isotopic spin and muons which will cause further simulation of phagocytosis and opsonic action," spoke the third voice.

"I don't understand any of this. I'm a philosopher, not a physicist!"

"Quite simply, then, David," said a fourth voice, "by such a system you will wipe out the firing systems and the memories of every weapon in the world. The secondary simulation of the phagocytosis and opsonic action we speak of means that your scientific device, when activated, will cause all armed satellites to neutralize the next, harmlessly, in the order of their position to one another."

"And that," said Adam, "is why I wouldn't invest in computers or the technology market, if I were you!"

"It will wipe out Microsoft, too," said Mursilis,

"like Babylon!"

"Not to mention money-lenders in their temples," said Adam, "along with a few program traders, CEO's secret accounts, lawyers, and insurance companies."

"That is humane," I said.

"Good! We all agree to that!" said everyone.

"Good!" I said.

"Now, David, just stay seated on the lovely bench from the Monticello River Park and close your eyes once more."

"No Diet Pepsi, Adam?"

"None needed, dear boy!"

The trip was short.

"This is for you, David, and our host, Big Lake!"

"My God, Adam! It is a missile! A colossal, huge, monstrous missile!"

"No, David, it is not monstrous. It is beautiful. It is our gift to humanity, since they rejected common sense and knowledge."

"But, what am I to do?"

"You simply push the buttons in the sequence we have provided you," he said, handing me a clay tablet, inscribed in Alkdian cuneiform writing.

And, I could read it!"

Then, Atum was gone. There was the slight smell of garlic, rosewood, olive oil, feda cheese and burning leather. Atum's black, high-top dress shoes smoked where he had stood before me.

I looked at the huge missile, far larger than any pictures I had ever seen of any missiles. The vapors were rising to the glass enclosed cage in which I sat.

I looked again at the firing order and reached forward to the panel.

The overhead door opened and with it the huge willow stump finally disappeared.

It was really quite simple.

OTZ, AK
1964-1965

SHEE FISH BY THE SLED LOAD — KOTZEBUE, ALASKA   178 FHW

Warren Tiffany, the Bureau of Indian Affairs School Director met us at the airport in Kotzebue, Alaska when we arrived from Anchorage by Wein Airlines MC (Mostly Cargo) on October 7, 1964. The official limousine was a 3/4 ton surplus truck of World War II vintage.

The two late-hire teachers the Bureau got in the shipment were uniquely suited to the school. Norma, my wife of just two months, was formerly a Reeve Aleutian ticket clerk, stewardess, baggage handler, cocktail waitress and a Ladd Air Base teacher. I was an entertainment specialist for the Department of Defense who had spent the past ten years in the Pacific and Hollywood auditioning talent and packaging shows for the USO. I hadn't taught in eons. If it was our idea to get away to start our marriage we had either come to the right place or gone to the end of the earth. And, along with us was our unique companion, Mishka, a dignified lady cat from the alleys in Anchorage.

Mr. Tiffany took us on a tour of Kotzebue, down the only road, carefully pointing out the commercial highlights, i.e., Hanson's General Store, Miner's Bank, Ferguson's Theater and Store, Rottman's Store and Bonnie's Bubble Room, the local Laundromat. All of these and everything else became critical to us in the months and years to come.

As we drove down Front Street we were suddenly told that the government quarters were full and we would be billeted in a Quonset hut with a wooden shed on the side only a mile from the school. However, on the up side, it was all ready for occupancy and only lacked a few items and conveniences. He must have felt like a devious real estate agent. He reluctantly took us there.

As he unlocked the door we began to notice a few of the shortcomings such as missing hinges, a definite lack of weather stripping and boards in the flooring. Lesser optimists might have considered these to be omens, but since this was only the doorway, we were undaunted.

The overhead light bulb worked and we were soon inside where the temperature was a comfortable 40 ° F., courtesy of a large oil burning 50,000 BTU space heater, an apparent age-mate of the surplus truck.

A Mr. Hunnicutt, said Mr. Tiffany, would help us with fuel oil and occasional tins of hospital distilled water. Cooking was to be done on a fuel oil-fired iron stove; the water heater coil running around some-where inside it and the water itself from ice melted in a 55 gallon drum with a gravity feed to sink and shower with a magic blending device for hot and cold, called a faucet or faucets, depending on which was working. We eventually learned to chop off the yellow portions of our ice supply before bringing it inside to melt. It, the ice, was left at the doorstep where any, or many loose huskies, could irrigate it as they passed by.

The shower Mr. Tiffany spoke of was a closet in an unheatable spare room, which without a curtain, also housed the honey-bucket and furnished everything in the building with the constant odor of Pinesol. We were to be regularly reminded that this was an expensive commodity in short supply and that the collection of garbage and night soil was on a mysterious schedule known only to a Mr. Green.

As it turned out, Mr. Green and his wife soon became our friends and visitors who taught us native drumming and dancing of classic proportion. The dancing and drumming classes usually started shortly after a couple of cups of the apple-jack I was able to brew in a butter barrel behind the cooking stove. We were consequently never to want for lack of expeditious garbage and honey-bucket service. The frequency of the collection was a mystery to the other teachers who never hesitated to ask how we rated so high.

To return to our indoctrination, our immediate despair was what Mr. Tiffany so proudly showed us in the Quonset hut storeroom. Besides two refrigerators which didn't work was a huge pile of supplies and foodstuffs called a "Standard Order." This was meant to see a family of 40 through at least five years of isolation where laundry starch, mothballs, fly paper and candied yams would be especially hard to come by.

Our distress was threefold. First, we had specifically and emphatically refused to take a "Standard Order" in Minnesota, Seattle and Juneau. We would have refused it an Anchorage, too, had we been given the opportunity. Second, we were charged $1,800 (reads $5,000 in 2002) for it to be repaid from our pay checks @ $300 per month, and third, the U.S. Government interest rate was 10% as compared to any bank's 3%. Freedom of choice was not one of the BIA's strongest point, nor was their willingness to admit error(s) in anything/everything they did!

Mr. Tiffany apologized as best he could; his hands were tied; it certainly wasn't his responsibility, and there was nothing we could do about it. He left rather quickly after that exchange, saying we were to report for teaching duties and class assignment at the start of the second shift at 11:30 am the next day. The first shift in the over-crowded school, he said, started at 7:30.

There was warm water by then, having lit the cook stove upon arrival and the sink faucet spurted fish scales and an odor strong enough to attract Mishka to the kitchen area, thus suspending her inspection tour of the rest of the "house." It was simple enough to figure out. The ice was from a nearby fresh water ice lake and the melting had produced results that even the Clorox additive hadn't nearly defeated.

We found our bedding, electric blanket, and towels in one of the boxes Jim Ennis had built for us and that we had air shipped, and, praise God, some non-Standard Order food items we had packed for emergencies.

There was one secret weapon we possessed living in Kotzbue and that was Leon and Vivian Shellabarger. Leon was a standout first cabin bush pilot and Viv and Norma had worked together for Alaska Airlines. We quickly made out presence known by walking into their hotel one block from our Quonset. They were surprised and Vivian was delighted to see her old workmate. Nothing would do but to sit down to their roast moose dinner, which we did without hesitation.

The next day we reported in and got a tour of the school facilities, conducted by Chris and Cleo Crawford, two magnificent people and teachers with rare qualifications for BIA employees, of masters degrees. We were greeted with kindness and curiosity by the off-duty staff in the combination teacher's lounge and work room, a 10 x 12 foot room for 26 people.

Our late hiring had created the opportunity for a mild conspiracy which resulted in the creation of a class of 34 fourth graders, age 10-15 years, for Norma. I was the recipient of a roster of 44 students, the result of a survey of students who had quit their BIA boarding schools in Edgecombe, Wrangle, and Copper Valley and returned to Kotzebue. Additional students for the Directed Studies program I was to teach would come from the adult community and what I didn't teach I could match up with the correspondence branches of University of Nebraska and the University of California.

During our second week, while walking home from school, we heard the crying of a dog. We found two husky puppies bleeding and hiding in the corner of an abandoned hut. We carried the pitiful little creatures home, washed their wounds and bandaged them with a torn up BIA bedsheet. Being an ex-medic

I had found a way of stopping the bleeding and you've never seen such a sight. I really think Eric got well because he felt so sorry for me.

The owner found himself with two too many pups and had hit them with and ax or hatchet, but quite obviously hadn't killed them. Their appetites were ravenous, their recovery nearly complete. We named them Eric and Evan and you could tell Evan from Eric because Eric was just a trifle cock-eyed from his encounter with elimination. Two more beautiful, loyal, playful dogs never existed above the Arctic Circle.

Over the next several weeks the pace picked up along with the coastal storms, winter wind, blizzards and 70 mile per hour winds. Our downtown location afforded us the benefit and joy of the world's first big screen television. Our shack windows faced the back of Ferguson's Store/Pool Hall/Theater on the left and on the right we had the Alaska National Guard Armory which served that function and many community based activities, games, and dances, for which the people turned out in great and celebratory numbers. Turning out our lights in the kitchen-dining room-living room combination provided us with a perfect view of those historic events and happenings. Lights out also prevented strangers from mistaking our house for the Armory at times of imbibition.

Mishka had created an unending source of wonder for the children as she was the only cat in the village with the exception of Annie Schaeffer's white Tom. Mishka was facing a crisis and we rented him for $10, plus room, board. and litter.

Under the terms of the isolated location for the Tom, he was totally lacking in experience. He was, to be honest, frightened by our amorous cat and it took six days of Mishka's most ardent and seductive snake dance to coax him from behind the two worthless refrigerators in the store room where we had cleverly put them together. She eventually won his confidence and, appealing to his masculine instincts, the inevitable occurred.

As the winter grew deeper and deeper, the walk to the school became more uncomfortable. We bought reindeer furs from the Bureau Agency and the dental floss to sew them came from Hanson's Store. Grandma Goodwin cut them for parkas and mittens along with caribou fur for the legs of the mukluks and walrus hide for the bottoms. They were wonderfully crafted and fit perfectly. The only problem was that the reindeer tanning was as yet unperfected and we left a trail of hair wherever we went and they were so warm we were compelled to stop and "breathe" the parkas by opening the top, then flapping the front to let the warm, damp air out.

As the Christmas holidays approached, Norma began a project of drawing and painting our greeting cards and I made an effort at the verses. She also planned a Christmas program and taught her class to sing Tannenbaum and Silent Night in German. Despite the fact that there wasn't much of a German community, two besides ourselves, everyone in the village seemed to enjoy it greatly and the parents were extremely proud of their suddenly tri-lingual children!

This was also the time to get the Sears, Roebuck, Montgomery Ward and Northern Commercial catalogues. We bought each other fancy ski pants, Nordic ski sweaters, apree slippers and though we lacked skills, hills and skies, we looked convincing and made frequent complimentary remarks to each other. Our rationale was not rational, but by golly, we were chic.

In February we had a weekend storm with strong Northwestern winds. By Monday the snow was rooftop level in many places. We gave ourselves a lot of lead time to get to school, but with the darkness down and the familiar landmarks buried, we kept walking into things such as poles and sharp-edged drifts.

Norma's mukluk came untied and she had to put an arm around my leg to keep from being blown across the ice sheet between George Francis's house and the school as she knelt to tie it. Only then did we realize we were surrounded by dog packs made up from teams unsnapped from their chains so they wouldn't be buried in the drifting snow. Their blue eyes were all around us, but they were only interested in our two pups, Evan and Eric. who were doing their very best to get as close to us as possible to avoid unwanted introduction and cold-nosing.

We arrived, dogs and all, at the door and to the surprise of the Principal, Charlie Richmond, who had secretly, at least to us, called off school for the first and only time that year. Norma found someone to hand on to that was headed toward town and I spent an hour with Doug Sheldon, the head custodian, bore sighting my .375 H & H Winchester 70 mount magnum I had left locked up in the downstairs supply room. (I was vastly over-gunned, even for moose!)

About noon when it was fairly light I started back and only got as far as the ice sheet outside the school's front door. I was sent sailing, parka over keister on the glass-like playground. The 70 mph wind skidded me into Schaeffer's dog keep about 50 yards away. Fortunately, I think, they had also let their dogs loose to seek shelter for survival.

About that time I really began to understand why the Eskimos laughed at us for being so crazy as to be out in such weather. There were stories of the Public Health Service nurse who had been lost out on the Sound for an entire day when she became confused on her way between the hospital and her nearby quarters. In the same story it was told that she found her way to a bayside house after sixteen terrifying hours. She had to be evacuated, not for injuries or from the freezing temperatures, but from the terror of the ordeal.

And, there were others about mothers whose babies had slipped from beneath the backs of their parkas where they were traditionally carried, and into the freezing snow and wind. Terrible tragedies occurred each year, and though not frequent, certainly often enough and real enough to confirm the belief that the Arctic is a killer and a moment or two of carelessness could mean instant death or near fatal consequences at the very least. We had been lucky, thus far.

Lady Luck, however, is a fickle fatalist. Yet another February storm, the third major that month, caught us on a Friday morning and by the time we finished our late shift and got back to our quarters, the wind had torn loose the outer door and jam-packed the entryway with drifted snow and hence into our wanigan. I was able to dig our way in with the help of the shovel we had learned to keep tied above the door. Yet ahead, were some interesting discoveries and some latent adventures.

We had noticed before going in that our three chimneys had been blown down and once inside, the worst result. The space heater and the cook stove had died. The good news, and there always is, was that the little space heater was functioning bravely and tirelessly. The snow had blown in between the roof and the ceiling and was now melting at a fairly nice clip. Not quite like Victoria Falls, but a good imitation of Minnehaha would describe it best. And, of course, it was re-freezing when it hit the linoleum floor. A veritable skating rink.

The excitement of all the activities was just too much for Mishka who decided it was time to increase the kitty population in Kotzebue by five, a neat 250% gain which is not bad for an Anchorage girl with a virginal Tom. Always one for comfort, she selected an open dresser drawer and insisted on Norma's presence as honorary midwife. If Norma even looked like she was leaving her, Mishka would start to laboriously (yes, laboriously) start to climb out of her delivery room, nee sock drawer.

Before restarting the space heaters, two things had to be done. The chimneys had to be re-attached and the soot removed from the fire pots inside the ancient Trojans. Getting to the chimneys was the easiest thing we did that day. One simply had to walk up the drift incline until you got to them. The cleaning of the fire chambers was a different matter. The wet, or so one would assume, soot would be subjected to the business end of the old Hoover vacuum cleaner and Voíla, the offending soot would be slurped and sucked into the bag. Things were going along swimmingly, so to speak, when it was discovered too late that the soot being vacuumed was anything but wet. It was, miracle of miracles, very much alive and burning.

The whole process ended with in a great cloud of soot and smoke as the bag burned through and blew out over everything north of Nome and south of Point Hope.

My being a black belt curser helped somewhat. Eventually, the night of cleaning and cat delivery and cold and soot and frozen floors and half-cooked moose meat stew boiling on the top of our courageous little space heater got to us. We simply looked at each other and staggered to the old delapitated couch and sat, laughing like idiots. Our two beautiful pups, Eric and Evan, ran in circles, slipping and sliding and yelping on the icy linoleum.

But we should have known the gods of the Arctic

were not through with us. After a belt of apple jack and an hour of trying to chew the moose meat and near raw carrots and potatoes and onions, we were more than ready for bed. Checking on the lovely kittens, we put the dogs out and headed for bed and the warmth and comfort of the trusty Westinghouse electric blanket. About an hour later, at 11:30 PM, things were becoming noticeably warmer, and it well should be the case. There was the unmistakable odor of an electrical fire. Right between us, even in the dark, we saw a burning hole spreading from a fatal insulator break. At that point it wasn't funny anymore. We unplugged the blanket and pitched it out the doorless Wanigan and onto the snow drifts. Semper Fidelis! Viva BIA! We pulled out the L. L. Beam sleeping bags and said goodnight once again.

Then, too, there was the democracy of the separation of church and state functioning at its highest level. Every six months the citizens of Kotzebue exercised their inalienable rights to vote for booze or no booze. And, it seems, the split between the wets and the dry contingent seemed to even itself out in just the manner to accommodate both sides. Every six months the town went dry and surely as the seldom seen sun would rise, the next election shut down the municipal liquor store. This gave the needed shot to the nascent bootleg business and the bamboo booze bombers made their runs to Fairbanks and Anchorage, turning a neat profit with each turn of the props. All this during the six months of supposed lawful sobriety.

At night the dogs on their chains in their keeps sang to the moon and each other, lulling us to sleep and during the day the marvelous originality and brightness of the school children encouraged us. Mr. Hanson, the fifth grade teacher collected and saved some of their writing. It showed just how descriptive the English language can be to one who has truly mastered it in a display of unusual expressions and superior verbal management, to wit:

"We cannot kill water because it is a liquid and has no air in it."

"He's got no muscle, but he is tougher."

"The dog is going crazy for the last time."

"My mind says go to school so I won't grow up to be a dumb girl."

"We had smashed potatoes for lunch."

"My aunt, her name is Easter, knows how to skin up a polar bear real good. She lives almost two house away from us."

"I don't like his fat hand, and I beat his head into the floor last night. His head will be so black. (Kenny beat up his big brother Charlie Tikik.)"

"In the desert people's tungs get fat and swell up like a bowloon."

"We had polar bear paws for supper."

"My puppy was freezing. I brought it inside and melted it."

" A pig ran around naked."

"If you go down there you're half way to Hell."

"They shot some peasants and other birds."

There was always something good, something

social, going on. A pot luck dinner at the school, our basketball games at the Air Force AC&W site, where grew the world's smallest National Forest of one transplanted pine tree and the inevitable caribou fever when they started moving south from the Brooks Range. There were also the seal hunts, the shee fishing and excitement of sighting the beluga whales. In the multi-purpose room of the school was the site of after school and evening games and once a visiting coed volleyball team made of townspeople and teachers from Noatak came to town for a week-end 12 game volleyball tournament. They whipped our butts.

Most of all, there were the children and the dogs. Erik's attention span had been somewhat altered by his injuries, bless his darling personality. He would get so excited and play and play until he forgot what he was so happy about. At that instant he would run to one of us or both of us to get recharged with joy and resume his romp with Evan or Mishka or the kittens who would arch up and hiss and defiantly put everything back in its proper perspective. Mishka, her Majesty, and the kittens had a inherent sense of dignity and morality which was not to be disturbed by mere huskies, despite their own rather question-able, engineered genetics.

Visiting dogs, those who came to our door each day, always had the most startled look when Mishka was outside in better weather. They, of course, had never seen such a creature in all their sled-pulling days, and cocked their heads this way and then that in unabashed curiosity and tentative uncertainty. Should they not attack such a strange, furry, feisty beast?

Mishka could sense their insecurity and made bold moves toward them, spitting and hissing while the mighty huskies backed off or simply shed any vestige of superiority and suddenly remembered they were needed at home.

And then we lost Erik and Evan, now so large and strong. The former master decided he wanted them back and he took them. The vacuum was a black hole, not just for us, but for a bewildered Mishka and her rollicking kittens, all of whom waited at the door for them to hopefully come in with us for a romp after school. We saw the dogs only rarely after that, but when we did they came at us like wild hyenas and we tousled each other furiously and joyously whenever we did, never forgetting for both sides to charge at each other full bore, yapping and yipping from both sides in the process of love and recognition.

Frank Gold, a teacher from New York and now Dr. Frank Gold in Bangkok, had a magnificent Kobuk husky of extraordinary size named Septuk. He loved that dog, but there was just no way he could keep him in the duplex where he and his wife Suzanne and two children lived or quiet him outside. We bought him from Frank and he was regal and beautiful, but not the void-filler we had with Erik and Evan. Some people are made for certain dogs, but we were not satisfactory to Septuk. A young minister in town needed a good dog and with Frank's parental approval, we gave him to the preacher. Both lived happy and productive lives thereafter.

Our house was located, as was said, next to the National Guard Armory and that was a source of great entertainment. On the other side of the Quonset was the village library. We visited it every time it was open. It had an almost unheard of collection and it was over-run since there was no television then and Armed Forces Radio from Nome must have been pointed toward Bethel to the south. The German-made Telefunken radio I had bought in England ten years previously and shipped to Alaska had a fine record player and that was a blessing. There were 33 rpm record albums available at the library as well as the books. The Nome Air Base had just closed and the very clever and diligent maneuvering by Mr. and Mrs. Johnson and their daughter, Debbie of the FAA brought the entire excellent collection of books and records to the school and the Municipal Library. Strangely, years later while teaching at Park Rapids, I mentioned that story to Andy McCarthy, the high school's tremendously popular and successful foot-ball coach. He had been in Special Services at the base in Nome and had in charge of the closing of said library there and the packing and shipping to us in Kotzebue. It was proper to put a face to the good fortune of our library.

When April finally got to the Arctic Norma de-cided there was a need for her class to travel. She had spent a year at the University of Hawaii and she taught the children all about the whole wonderful business. They made grass skirts from green crepe paper, ukeles from cardboard, paper flower leis and learned to sing Hawaiian songs. Norma's piano was frightfully out of tune, but no one cared. They democratically elected the pilot, co-pilot, engineer, and the flight attendants. They made tickets, bought and sold the tickets and formed the forward cabin, first class and tourist class with their chairs.

The excitement grew as she put on an old Johnny Ho record, turned on the school vacuum cleaner for jet engine sound effects, and they were off! They flew to Anchorage, changed planes and arrived, on time, in Honolulu. They danced the hula as Norma had taught them, robustly sang the Hawaiian War Chant and other selections they had learned, had a great luau, thanks to Norma's ingenuity, hard work and some very early ordered fresh pineapple, courtesy of Rottman's store. The kids were sensational and the parents went wild. Never before, and perhaps not since, has/had such a flight originated in the fourth grade of an Alaskan native school.

When spring came, I received word that I had been awarded an NDEA grant to attend the University of Alaska, Fairbanks to study history and Norma learned that she had been accepted there for further art study. We left Kotzebue for 12 weeks and found a lovely log cabin on Minnie Street in which to live with Mishka and the runt of her litter, Fritz. Fritz was a bit slow and killed the same shrew, daily discarded in the back yard, over and over until the poor thing simply fell apart. It was all very exciting and the University life was challenging, stimulating and fun.

Despite all that, on August 28, 1965 we were the first ones on the plane taking us home with our hold baggage marked OTZ. The wee Volkswagon we had bought for our use while in Fairbanks followed on another plane, filled with all the wonderful food items we could pack in it. They were, as you have probably already guessed, a far cry from the dreaded Standard Food Order. It was never a question about the inappropriateness of having a car in a village with only one road. It was a lovely little car and when it arrived three days later the Wein Airline's agent called the school to tell us that our food order had arrived wrapped up in a Volks bug!

May Day In September

May Day in September was first published in the
June, 1984 issue of the Alaska magazine

Wally had plans to run his dogs that winter and that seemed like a good idea at the time, but 14 dogs eat a lot and if you want them trail-ready, you've got to feed them pretty well. A caribou a day is a little too much, but a half of one isn't. If it's too early for the migratory herds to appear and Kotzebue Sound or Hookham Inlet aren't frozen, they're great for fish. This is what you want for your dogs. It thickens the coat without fattening them.

If I helped Wally get fish, he'd let me freeze with him on the trail with the dogs.

It was 1966. Wally Blasingame and I, together with our wives, were teaching Eskimo children in the Bureau of Indian Affairs School in Kotzebue, Alaska. Kotzebue was best known then as the polar bear hunting capital of the world and there were enough trophies taken and records established to prove that. What was a lesser known fact was that the people there in Kotzebue owned more boats and motors and did more subsistence hunting and fishing than any other town north of the Arctic Circle. There are still a goodly number of aircraft based there, many of them owned by guides and outfitters, but a lot of people just like to tool around in their spare time.

Back in those days the only car in town was a '63 Volks Beetle Norma, my wife, and I had flown in after summer school at the University of Alaska, Fairbanks. There were no roads in or out, and there still aren't, but now the population has grown from 800 to 5,000 and there are hundreds of cars. A nice lady at the school told me in the summer of '01 that our old Volks was still there, but it wasn't functioning at the moment. There's even a bus service and two taxi companies. I guess that's another phase of progress.

The bay off Kotzebue, between the town and the many channeled approach to the Noatak River, is usually quite calm in September, but the tides and subsequent shifting mud shoals and sand bars can be tricky to navigate. Wally and I were confident that the weather would hold that week-end and we could make a run up the river and locate some likely sloughs to set out nets. It beat the heck out of open-water netting and thus we made our plans for an early morning start.

That Friday afternoon after school we checked the equipment, tested the outboards and winched the boat ashore with a truck to tip it on it's side and check the keel. Wally had just built the boat the summer before and it was a sturdy 25-footer with an open cabin. Satisfied, we were ready and the boat went back at anchor.

Shortly after 6 am that Saturday morning I hauled the bow anchor aboard as Wally started the two outboards; one a 35 hp Evinrude and the other a 40 hp Johnson. He made a sweeping turn at half power and we headed toward Lockhart Point, the local landmark and our point of departure for the Noatak River.

The sun was bright and it soon warmed to 38° degrees Fahrenheit. The bay was choppy and the wind was light as the miles slapped the starboard bow behind the heavy hum of the twin motors. Coffee from the cabin primus tasted good as we enjoyed the Arctic morning. Wally tested the single engine speed as I found my way through the nets.

Across the bay and into the spray we went. An occasional exchange with muddy water stretching for the sea toward the land's end of Kotzebue spit was felt. You could hear the change in the props as they cut through the heavy silt-laden water. Fat geese rose to our right as we drummed the sleepers from the flats of the Little Noatak.

There is little to tell a person whether or not you are in this week's or last week's channel. A Cheechako is wise to wait for someone else to make the run for the river before you do. Our luck was with us as we hit out on the first shot. We weren't Cheechakos and suffered no sheared pins, bent props, nor did we stove in the bottom.

The rest of the channel search was a piece of cake with the exception of a few instances where the banks were cut away on both sides and our guess could have gone wrong and rung us up.

A few of our native friends were dog fishing, coffee cans upside down at their shelters. Nothing was obviously tearing at their nets. There was a sullen and sour look of those who see the days shortening and the winds rising with the howl of their teams.

We ate a casual lunch as we tied up to a willow along the bank. The routine of over-boiled coffee, pilot bread and surplus World War II orange marmalade while we transferred gas from auxiliary to main tanks. The barometer was tight on 29.68 and the wind was low and steady, slowly shifting from west to north by northwest. Ducks on the pot hole to our left could be seen. Three fell to my 12 gauge double during a wide open walk past the bog at just 40 yards.

We cast off and made the final ten mile run and searched the adjoining sloughs. Nothing! Absolutely nothing was moving through the water either in or out. Not a blamed thing to get our nets wet and tangled over.

The skies continued to darken as the wind completed its sweep toward the northwest. The cold began to assert itself as the winds touched the hills already blue-white with snow and ice.

The return down the river was hastened by the evening tide and the desire to hit the gap before the winds arrived. A tie wouldn't be good enough, with a 3/4 following 15-knot breeze.

The oncoming darkness began to obscure the telltale outline of the mud flat as we shot from the mouth at nearly 30 knots.

The channel was gone! At full bore, one engine halted as the other slowed to a painful, laboring, "chunk-chunk-chunk." By some miracle we kept moving. We both slid over the side and up to our hips in the water, sinking into the mud, pushing, shoving and swearing, we were free at last to roll back aboard over the gunwale. Wally steered manually while on his knees in the stern, momentarily too whipped to move forward to the wheel.

The flashing airport light led us north of Lockhart Point where we hugged the shore and eased past the short sheltering breakwater and into the lagoon where we must tie up for the night. Even the 1 3/4 ton bulk of the 25 footer takes a terrific beating if you anchor on the bay side when the wind comes across the inlet. It was so calm in the lagoon, but even so, the spray was freezing on our hands and faces as we unloaded and anchored out again.

We waded ashore and started the mile and one half mile forced march to coffee and warmed over Saturday supper. An early bed and a before dawn jump-off was agreeable for Sunday if the weather, never predictable in the Arctic, was cooperative. Both of us probably hoped it wouldn't be so we could sleep in.

That thought disappeared quickly on the following morning. It was a beautiful sunrise with an almost spring-like five-knot puffer which lifted the smells of dogs, fish and breakfast smoke across the village. Wally and I had both jammed extra rations into our watertight kits before we met at the government road to walk to the boat. We knew it would be a long, hard day of netting, but one good day of netting can get enough fish to last the dogs until the huge caribou herds started their annual trek south from above the Brooks Range.

We were at the mouth of the lagoon and we looked toward the far mountains and trailed to starboard where the sun glistened on calm water beyond Lockhart and the shallows south of the gap. It was rolling smoothly and it should be good past Pike's Spit and down Hookham Inlet by settler John

Nelson's summer cabin.

Without comment we nodded to each other, knowing that our hunting days up the river were over. We must go for broke in the open water of Hookham. Wally started the engines and nudged both throttles and the heavy boat banged its way grudgingly into the smooth swells.

As we cleared Lockhart Point we decided to make a test seining between the mouth of the Noatak and Pikes Point, then beach at Pete Saterly's cabin for lunch. It was calm and warm as we slowly cruised across the bay. We re-tied the nets for the test run. We took no notice of the increasing north wind and the swells giving way to chopping waves.

We spotted a whaleboat off to our port. She was closing on us on an interception course. We weren't surprised to see the boat, but we did wonder why at the reason for the speed at which she was traveling. We soon recognized the young native men from the village as they hauled up on the leeward side. They had noticed us in the no-where-ness of the area and thought we might be in trouble because we were obviously nearly dead in the water. They helpfully pointed out where they had pulled 20 odd sacks of shee fish close by. They had only that morning taken up their nets for the winter and they suggested we try the same spot. We thanked them and they gunned their engines and were out of sight on their way home in minutes.

With that bit of encouraging news we changed our minds about the test netting. We began to re-rig for a deep setting. The wind kept rising, but it was only a mid-morning freshner. It was too bright and early to take the day seriously. We worked hard, but not very effectively as we were using two different sized and weighted nets; a price one pays eventually for buying used nets. The whole line would have to be re-done. That's hard enough on dry land where you can spread out, but a genuine problem when in the confines of a boat cluttered with Blazo cans, a 55 gallon fuel drum, and a junk-yard of other gear.

Wally selected a spot about a mile and a half out from the mouth of the Little Noatak. It was three miles distant from the land mass southeast of Pike's Spit. There was the usual problem of holding into the tide drift against the wind. I was now in the bow and got the anchor in at just about the right spot when I was almost slapped down on the forward deck by a combination of wave action and just plain stupid carelessness. I had let go of the lifeline between the hawse and the cabin roof.

Wally was busy with the nets while I made my way back to the well deck. Then we lurched again.

All at once I saw them.

It was the waves. They must have started well out to sea, more than 25 miles away. I'd seen them before while in the Navy in the Pacific in World War II. Every fifth wave was a monster. By the time they got to the gap between the Little Noatak and Pike's Spit they were really rolling. They slammed underneath us in the shallow water and brought the bottom of the bay up when they hurtled into our side.

We had the net out about 40 feet to our port stern, securely bent on a Blazo can float and well anchored by hooks. We were knocked into the trough and then held by the anchor at the bow and the fouled net which ripped into the dead engines.

Without saying anything we knew what we had to do. The waves were beating the hell out of us from the stern. Water was pouring over the transom. We had to heave the rest of the nets and floats overboard and clear the engines. There wasn't time to try to cut the wire rope on the top of the nets.

Wally's tremendous strength was shredding the net where it had strangled the engines. I hit the forward deck in one flat dive to shove and pull myself to the anchor line and try to pull us bow-to. I was on my back with my feet braced as best I could on the cabin cowling, pulling my guts out when I heard Wally yell.

"To hell with the anchor!"

He was passing up a life preserver to go with my Mae West jacket. To my surprise, I was standing at an unbelievable angle as I threw the kapok preserver over my head. I watched the engines go out of sight in surreal slow-motion. They didn't come back up. We were swamped.

It was all so rehearsed! I found myself moving quickly to stand on the lee gunwale, waiting for the inevitable turn from windward. A wild kick from a wave and the boat rolled the wrong way! We did a handstand, then dropped to the side. We were instantly jerked violently up and I lost hold. I went under, instinctively throwing my arms outward to keep the lifejacket from being torn over my head.

Wally held and rode it, pulling himself back on the ridge of the gunwale while my cotton gloved hands made a grab for the side. The boat filled and turned the way she was supposed to. Wally calmly timed the turn and went up the side and over the bottom to the keel, not once getting above his hip boots! I remember marveling at this, thinking he must have been practicing for this event.

Strangely there was no panic, no wild struggle, no frantic effort, as the waves, now three feet high, crashed across the top.

I don't know how much my two size nine hip boots full of water weighed, but with my 165 pounds, a slippery fibre-glass boat bottom and high waves, it was quite a chore to pull myself up. Wally, bracing himself as best he could on the keel and by interlocking grip of hands and a slow draw I pulled and was pulled to the three by five foot space, which in better weather would have been above water.

After a moment's rest, Wally reached inside his jacket and produced dry cigarettes and a lighter. We looked at each other and smiled. At that point it would not have surprised me if he had asked what I'd like to drink before dinner.

We talked about our situation, It was very bad, but not hopeless. There was little possibility that another boat would be out. The sudden ferocity of the surface wind would have driven any others to shelter or shore very quickly.

We were too far down in the sea to be spotted by Pete Saterly. He was the only settler sure to be around within 20 miles. Pete watched all the coast as best he could, but he had hunting and fishing to do. Any help from that quarter had to be realistically dismissed.

Having reached one conclusion, we pulled the hoods of our waterproof parkas over the tops of our heads, locked the zippers and put our hands inside into our armpits and squeezed. We squeezed and strained again and again to increase the circulation. Looking at each other we saw no ice was forming on either of us as the waves washed over and submerged us to our navels as we sat straddling the keel. A little spotted seal watched us from 20 yards away, disappearing and reappearing between the waves as they snapped up and over us.

There hadn't been a plane in the sky all day, and being Sunday there was little chance that any bush pilots would be out, particularly as darkness would soon be on us. We were certain neither of us could survive a three mile swim against the tide to the shore at Pike's Spit. We also knew we couldn't swim toward the tundra and mud flats at the mouth of the Little Noatak and there would be nothing there to help us should we make it. The current would be just too strong to last the mile and one half effort.

By 2:30 PM the sky was darkening as it had on Saturday. It could only get worse. That being the case and since I was already soaked and freezing, I asked Wally to pass me his knife. Mine was gone from its sheath.

I told him I was going to drop over the bow and cut the anchor rope. We discussed this slowly. We both knew that decision could be fatal. The wind was blowing opposite to the run of the tide. Would the boat go with the waves and run us deeper into the open end of the bay and hopefully to an eventual landing on the flats? Loosened, the tide could carry us past the gap and on out to the treacherous open sea. Without the tremendous pull against the anchor we might be turned upright again and onto a mud shoal and beaten apart by the waves.

We decided to remain as we were. We could, possibly, survive until morning when a search would surely be mounted.

It was now 3:15! We had been in the water since 10:50 am. The cold was growing and the sky had turned an ugly gunmetal gray with the unmistakable cloud mass of a cold front moving steadily toward us from the northwest.

I was shaking spasmodically as I watched the darkness deepen in the shadows forming at the side of the mountains to the southeast. Death was approaching.

Then, in a quiet voice against the wind, Wally said, "A plane."

A silver dot appeared backgrounded by that ominous bank of clouds forming the front.

It was a plane. It was following the shoreline at about 5,000 to 6,000 feet and three miles away.

Wally stood, nearly up to his knees in the waves

and waved his orange life jacket to lend some animation to our desolate watery wilderness.

The plane went by us. My thought was that we were doomed. In that moment it seemed all hope was disappearing.

Suddenly the right wing dipped and that beautiful aircraft nosed into a power-on glide directly at us. It was obvious he'd seen us and within a few moments the pilot waggled the wings and changed the prop pitch to signal that he understood the situation and would send help.

Taxiing toward the end of the Kotzebue airport runway for a bush run was Don Ferguson, the town mayor and son of the Arctic's famous Archie Ferguson. He had heard the call from Public Health Service's Dr. Compton, who was flying the plane from which Bill Jones, the hospital ambulance driver had spotted us. Don reacted instantly.

He barreled down the runway apologizing to his passengers for what he was about to do. When he reached the edge of the lagoon he shut down the engine, jumped out, ran to the float plane area and promptly "borrowed" Leon Sehllabarger's float plane. He kicked it over and took off toward the radioed position. Dick Heinz, the duty operator for the FAA notified all aircraft in the area to stand by, advised Joe Narcisco, the FAA station manager who in turn hit the alarm switch for Air Rescue 480 miles away in Anchorage to coordinate all efforts for rescue.

Within 10 minutes after spotting us Don was making a high, power-on approach to check the possibility of landing on the extremely dangerous, rough water. In 20 minutes there were 17 planes at various altitudes circling us and Don roared down into the waves. He had to take off again. It was just too rough.

Don corkscrewed Leon's plane up and came back down at wave top like a fighter plane on a strafing run. The waves were three to four feet high and boiling. He held it right at the top and stopped within 25 feet of us and bulled his way to the edge of the overturned boat, swinging sideways at the last second. Wally shouted for me to go and I grabbed a strut, stepped onto the left float and stiffly pulled myself through the open door of the plane's cabin. There was hardly a pause as Don gunned it and continued into the teeth of the wind. He headed directly toward Pike's Spit and landed in the shelter of that promontory in minutes.

Alertly, Pete Saterly had me in his cabin by the time Don had turned his bandit plan around. The right float was very nearly submerged, but he ripped it from the sea and repeated his fantastic performance, snatching Wally off the boat bottom. He flew Wally directly to the lagoon and turned the plane back to Leon who didn't even cut the engine, but instead headed down the lagoon on his way back to pick me up.

My feet were in a tub of hot water and I was wrapped in wool blankets and fed a hot brandy by Pete and his guest who had just returned for a visit to his friend after a 17 year absence.

Leon landed and brought his plane to near the door step. In borrowed clothes I climbed aboard and with the weather worsening by the minute we started to take off. All of a sudden the power slackened and Leon was forced to turn around. We couldn't make it. The right float was by now underwater. Leon made a radio call to have his famous "bamboo bomber" stoked up and sent out.

Within minutes, Warren Thompson, an FAA pilot came screaming over the top of the high point and dropped into the cove. I was stuffed in and we were off for Kotzebue. I was pleased to see that we were taking the land route home.

I asked Wendell Wassman, the Wein Alaska Airline Station Manager to drop me at Wally's place, which he did. I think we looked pretty good to each other.

Wally came with me across the way to our quarters to help me calm my nerves and help assure my wife. Norma was as calm as could be. Wisely, no one had awakened her from a pleasant Sunday afternoon nap during the rescue ordeal. As a matter of fact she didn't notice my strange clothes for fifteen minutes.

That evening I drove Wally and myself in that one-of-a-kind Volkswagen down Front Street to see Don Ferguson and Leon Shellabarger and to thank them again. Don was calmly selling tickets at his theater and Leon was home in his hotel with his wife and three children. They were talking about the coming polar bear season and a possible trip to Hawaii afterward.

The next morning, Monday, there was ice over the

sand bar. On that afternoon Wally got his boat righted and towed back to the lagoon without the loss of so much as a can of beans from the cabin. My double barrel still worked just fine.

You know, when John Nelson found the floats and our net, we had to laugh. He picked it for eight sacks of shee fish.

He then gave them to Wally for those dogs.

The Dead Do Stir

Charles Darwin Calvin sat in a deck chair on the balcony of his tenth floor apartment in Vina del Mar, Chile. He faced the sea under a floppy fatigue hat of white duck. It was 1984.

Calvin was 58 years old and in the intriguing process of slowly and deliberately killing himself. It was something he had contemplated at agonizing length. His was not the ordinary depression and swallow-your-gun type of self destruction. It was as well planned and organized as anything he had accomplished or thought of accomplishing since he was 20. Always methodical, Charles determined to make this dying effect last; not to suffer, but to atone, at least in his mind, his acts of wrong doing.

Charles had failed to adjust.

He smiled and shrugged his shoulders in resignation, reaching to his right and picked up the first of ten pills carefully arranged on the marble table. One for each half hour or so. His calculations were based on his body weight, tissue absorption capacity, age, heart, lungs and blood pressure, all taken into consideration by the Embassy physician who had given Charles the pills. Dr. William Warren had a lot more pills than the ten he had given Charles. The meticulous character of the intended dead man was an imperative issue. He had insisted that everything be right and that there would be no point at which he would simply gulp them all to end it quickly. It was to be done righteously, with only his own characteristic cerebral celebration and discipline.

There would be three pills to take at the last.

He stood beside the table, bent over it and took the first pill, drinking some bottled water from a heavy crystal glass. He counted the pills again and looked at his watch. It was 3 PM. That would make him dead by evening.

"Fine," he thought.

Dr. Warren had told him the pills would make him neither dizzy nor nauseous. He sat in the deck chair again and unconsciously checked his watch as the single half hour tone rang from the bell in the church next door and below him.

"'Like the river to the sea, I drained myself,'" he said. "I'm glad it's over." He pushed his white hat forward to the edge of his eyes, kicked off his sandals and let his chemistry blend with that of Dr. Warren's.

Keenly aware of his pace, Charles could sense the interior he was setting adrift. After fifteen minutes he was serene. He snorted to himself and his mind spun backward, upstream in his own River Urubamba to his beginning with the CIA. Thinking now of "Henry's Understanding" when Berryman had said, "A concentration upon now and here, suddenly, unlike Bach, and horribly unlike Bach, it occurred to me that one night instead of warm pajamas, I'd take off my clothes and cross the damp cold lawn and down the bluff into the terrible water and walk forever under it out toward the island."

Charles told himself the narcissism was not as dead as Berryman, who had preceded him into the river. Berryman's was the Mississippi River and an angry plunge off a bridge, not a naked walk under it

to an island. Charles was several steps behind
Berryman, but there were hours left in which the
rationalizing mind could cause him to stop amid the
slow strides on his own cold lawn.

He heard the clock striking from the sitting room,
behind the double French doors. He looked at his
wrist watch and heard the church bells again.

"They're not going to let me forget the time, those
goddam Catholics," he said in one of many anachro-
nisms.

He sipped more water as he took another of the
doctor's specials. He stood again and went to the
refrigerator and got ice in a thermos container, more
water and a bottle of beer. It was hotter than he
realized, or was it the drug? He hoped he wouldn't
become morose or morbid.

It was the same Mississippi River, as Berryman's
adopted death water, and the well-spring of Bly and
Veblin and Gus Young from which Charles had come.
Long ago he had assumed his manhood there and lost
a wandering wife somewhere since then.

Protestant ethic, puritan parents, purile puberty,
and now popping pills. Wandering wife, faithful,
loving wife, it was all the same. "Gone, old man, but
not forgotten," he whispered aloud.

Distorting now; a jock strap his mother had
bought for him because he was too shy to buy it at the
drug store. Ordered by mail at the dining room table
from Montgomery Ward in St. Paul. A five day wait.
An empty symbol of maturity. Plenty of time in
practice for sex and later in the games and the bus

rides home with the cheerleader's tits in his hand, bare and hot. Great in high school to say nothing of the summers on the sand bar with the vacationing city girls, so wise, so willing, so pretty.

How hard to stay a virgin lest you make a young girl pregnant or have to marry someone you didn't love and couldn't respect for what she'd helped you do.

And then, the big bang, not with a girl, but the war and lots of girls whose interpretation of patriotism was placed in their crotch.

God! How he had loved his country then and all the helpful people in it who made him proud to be in the service and bashful about his ribbons and his Silver Star and Purple Heart for Okinawa's Yanabaru and Yellow Beach Two.

Perforating belly wounds and peritonitis and piss and puss and oozing gangrene with yellow sulfa and atabrine atmospheres about their bodies, everywhere, yellow, dead or alive. "Yellow stinks, you can smell it coming up your leg," Cliff said before they used the meat saw about five inches above his left knee.

"Careful what you cut off," Cliff had warned from the blood-brown stretcher in the Iwo Jima sunlight. Not even a shadow from Sarabachi to grace the severed leg. Too sick to turn his head to puke, Cliff gurgled with it in his throat until Snowden cleared the half digested ham and egg C ration from his esophagus. And Snowden was black and Cliff was from Mississippi.

God, how Charles Darwin Calvin loved his country because she was strong and right.

/"And death shall have no dominion,/ Dead men naked they shall be one"/

Sweet home and the bands played on that Memorial Day. "Under the Double Eagle" and "Washington Post March" and Sousa's best by boys and girls who had seen it all in the newsreels, followed by the shrinking uniforms with fading division patches. Out to the graveyard and a volley of blanks from M-1's and Enfields by the boys from W.W.I.

/"Time enough to rot; / Toss overhead/ Your golden ball of blood/ Breathe against air/ Puffing the flight's flame to and fro,/ Nor drawing in your suction's kiss"/

The decision of what to take in college was a lot more difficult than taking orders from a ninety-day wonder. He got a job as a janitor's assistant and scrubbed woodwork in the gym in winter while the ball team he should be playing for was practicing. He spread fertilizer on the campus lawn in the spring while the other guys were making it with the girls lying around with their skirts never quite tucked under them. The girls saw him, though.

"Oh yes, Chuck, I've seen you around the campus. You work here, don't you?"

Or the studied but not-quite-making-the connection stare during class time. "I've seen you somewhere before." Never dreaming that anyone on the GI bill had to actually sweat to make it through.

Or in his evening's natural haunt on the main floor of the Library behind a stack of Mitau's political science books or Waldo Glock's geological monstrosities echoing the African Genesis.

"Chile's geostructure happened with surgical purity, cut by the knife of a fault and amputated by the Andes."

What a monumental view of afternoon blue!

But after the library closed at ten PM he could smell girls on the steps among the mosquito slaps and conversation with ceremonial words about coke or coffee, the ritual of beer and burgers. Standing one foot in bobby sox and saddle shoes on the step and one foot on the sidewalk, ready to turn toward and walk away with the best offer or the one they really wanted to be with and had been waiting so long to meet. The one from another town or from the city who would listen to dreams and ambitions and actually be interested in them instead of what they had between their legs.

Charles had been afraid. In this city too shy to be aloof, too proud to be phony. The girls wanted genuine people because, as always, they were more real and genuine than the bluffing, blustering little boys around them.

Yield not to temptation, either side, but yield they did, sorry lot, sorrier later that the hormones peaked while their better judgment had been lost on a blanket or in the back seat of a dinged up Ford with a channeled ass-end.

Which ultimately precedes which, the body or the behavior? Only the speechless, ever watching single cell knows. It is all inside, all of everything.

Charles grunted in approval of his recall of such a question and the answer to all questions.

From Machu Picchu to Huayna Picchu as Charles had done while on station. He traversed the saddle in between the heights of World War II and the lesser grandeur of the Korean Conflict. His college was over, the timing just right. They had left him alone from his discharge in 1945 until 1950. Four years to college and then another war.

At first roll of drums, Charles came out marching in favor of territorial integrity and was now a commissioned officer with the 9th Infantry Regiment of the 2nd Infantry Division.

Scaling a ladder over a sea wall at Inchon he caught a secondary fragment from an exploding mortar and found a small world around him when Snowden, now an officer too, smiled at him, pushed his helmet back and said, "Calvin, you sure as hell got a penchant for our collecting stations," then tagged him for a MASH and went on to another stretcher.

Snowden, dear Snowden, where are you today when I need you?

/"This autumn, I / Cannot find the road / That way: the things we must grasp, / The signs are gone, hidden by spring and fall leaving / A still sky here."/

Saddle up and move out. Tangent, cosign, ellipse, the trapezoidal doors and gates and niches without a curved wall save where walls must bend, the interlock of architecture and war.

"Gauze and body bags and ID tags," he sang as the morphine reached his brain and a medic with a mustache and vacant sleepless stare put his stretcher on one side of the little Bell helicopter.

/"Behold, thou are taken in they mischief,/ because thou art a bloodyman with horror loud / down from Heaven did I not then hear, / but sudden was received."/

A tingle started in Charles' left leg and he raised it, sliding his empty hand down the sweat between the slats of the chair and those old memories.

He saw the time and sat up to take his third pill. There was still ice in the thermos and he dug into it to pull out two cubes, plopped them into his glass and took a long drink. The perspiration around his hat band matted his dulling greyish hair and he wiped at his forehead with his right forearm. He held his hat above his head. Just then the late bronze bell bonged once then returned to rest.

There had been little to squander in the area of adoration for his physical return because everyone was gone. One year in combat and eight months at Fitsimmons General Hospital and he was just as anonymous as the dead ones.

He wondered now, how he had pondered the paradox of the simultaneous pursuit of two objectives: to keep the nature of his own personality from being understood and to master or predict and control the behavior of other human beings. All for the good of mankind, despite the incompatibility of the two.

He made a waving gesture from his balcony and Washington came steaming into view with its humidity and quiet quickness.

Fifteen pages to the application and an interview that he couldn't remember except that the interviewer was from Amherst, which later seemed terribly funny and incongruent.

"Would you come back on Thursday at 0930? We'll see what we can do for you." It was Monday.

Now, on Thursday there were three men, and a young lady with long, beautiful fingers in the room. All of them commented on his war record and all were interested in what he thought his area of expertise might be.

"I don't know, but I'm willing to serve in any way I can."

"Captain, could you kill again if you were certain the person was as much the enemy as his counterparts were in the Pacific and Korea?"

"I'm sure I could. It would be different, but I could. Yes."

"Yes," they nodded, "We're sure you could, too, no matter how odious."

"Thank you," said Charles.

"We shall be in touch. Please do not leave Washington until you hear from us. Voucher all expenses on these forms. Miss O'Gara will handle your file in the future. She's more or less in charge of you now. Oh, and she's from Minnesota, too. Minneapolis, isn't it Miss O'Gara?"

"St. Paul, Dr. Shoen."

"Right. Sorry." and the three men had left.

"St. Paul. I gathered as much."

"You're very perceptive."

"You're very beautiful, or shouldn't I say that to a superior?"

"You can and should, Captain."

"Thank you, Miss O'Gara."

"Casey."

"Casey?"

"Katherine. Casey."

"Oh, I'm Charles."

"Chuck?"

"No, Charles."

"Good. Charles"

"Dinner tonight?"

"Yes, please."

But Charles could not find his way back to the kind of truth he had once touched. He could not experience the ecstatic lightning let loose by love again.

/"I could only grasp a cluster of faces or masks / thrown down like rings of hollow gold, / like scarecrow clothes, daughters of rabid autumn / shaking the stunted tree of the frightening races"/

Charles Darwin Calvin, the stunted tree. Lost amidst the hardwoods of humanity.

/"What was man? In what layer of his humdrum / conversation, among his shops and sirens - - / in which of his metallic movements lived on / imperishably the quality of life?"/

Like the submarines at Rota in the summer of 1959 and the B-47's at Moron, poised on the line, too, at Torrejon and Sarragoza with the F-5A's in escort.

Waiting for the inevitable challenge of the vapor trails of Soviet reconnaissance photographing the deployment of the magnesium magnets below. The ant colony which pulled its atomic eggs closer and closer, all around the mountain home of mother Russia.

"The most mentally deranged people are surely those of us who see in others indications of the same insanity we cannot afford to see in ourselves," observed Charles, nodding at his new wisdom.

He removed his hand from under his leg and looked at the hair on the back of it, still with a golden color, neither dark nor gray.

"God, if there was but a Spain again."

Charles knew his connection to Spain, maybe always had, but more clearly in later years.

"The function of God's Iberia is more just and moral than political," he heard his voice say aloud.

In massive ignorance and poverty, Spain's people, progeny of empire eaters, glowed with the only real greatness, the greatness of loving reality. Reality after recognition has passed to successor, the victor. "To the victor belong the spoils? No," thought Charles, "Victory only corrupts and rots the winner. Spain has always been spiritual."

La Feria, flagellation, a fire cross, joy and self-denial. The purity of knowing, of being sure. Maybe that was what he felt, what he knew.

/"Lord, have mercy on my son: for he is lunatick, / and sore vexed: for ofttimes he falleth into the / fire and oft into water. / And he did evil, because he prepared not his / heart to seek the Lord."/

Another pill, another swallow of water. The beginning of the urge to get on with it. What the hell, it was ten stories down or six tablets to go, either way. Wavering indifference.

He owed himself these moments. There was no point in just hurrying it through. The step over the balcony and the wild plunge, the wind rushing by, muscles tearing at bones, breaking them before he ever hit, the slow-eyed sight of the building going past, the people below, the vertigo as he plummeted in a dive to destruction, hearing his own scream of primal agony. That was not for him.

But the death, the dust of dying filtered through. The maggot, the little death with fat wings laid siege to him and the ominous dwindling of his day was black. No hope was ever Charles' for the permanence of stone after the many lives he had consumed in dreams. A shred of worn rug beside the screen door at home, napping there through the heat, no rage now, no excuse to remain. All childish, he was sure.

"How many more tones of sound, how many bells to ring, drenched in shadows of the moment of memory do I seek?"

Charles remembered that he was there to die, not in tenderness, but, not in violence either. Just to die the way he wanted most to die, which was more than most could choose.

The admission into the arena of his death was paid in remembrance. He had no reason so question the living world, but his fear of death was obliterated by his conscience. Not that Charles felt personally responsible, but he could no longer go on blaming others or excusing them either. His guilt precluded the thought that it was all just a trembling accident germinated without thought or plot on this hothouse planet. Someone planned what he had done. Someone, somewhere, with devious, calculated, dishonesty and greed.

/"Lo, where in this whirlpool sheltered in bone, / only less whirlpool bone, envisaging, / a sixtieth of an ounce to every pint, / sugar to blood, of coma or convulsion / Or, coma and then furious convulsion"/

And Salvador Allende is dead, too.

The white heat of the sky's sun shone now on prominent places in the patchwork of his balcony dilemma. Charles Darwin Calvin shifted his features somewhat and sighed at the conscious return to the reality of the equator; the here and now. His presence of mind was like the blow flies on the corpses once in streets not far away. He feasted on the festering remains of recent historical fact.

"Gary Powers was crazy as hell, there at the end," he said and added the necessary, "but who flew the plane and who spilled his water on the floor of his cell every time someone opened the goddam door?"

"Not me," said Charles, "Not me!"

"But, Charles, you said he had to go on that April day."

"Gary liked the money!"

"Not that much."

"I know, I know! I'm buying the gas right now!"

/"Now / Say nay / Sir, no say, / Death to the yes

/ The yes to death, the yesman and the answer, / Should he who split his children with a cure / Have brotherless his sister on the hadsaw" /

"Yes, the dead do stir, I think it more and more. And I awakened part of them each time I placed another upon their heap," thought Charles.

The sharp tones of the sunlight shook his vision. A commitment of intensity which moved his mind to the bright sight of the Bay of Pigs.

For Jesu, his sleep was in the moontide; garbage, whose feet went eight yards home before he fell, eyes front, in the sand. We all believed. We all hoped, but not as much as the sixty souls of the other Spanish speaking turbines from the swamps of Florida.

John Fitzgerald Kennedy, son of Joseph, the baron of booze and recent of the Court of St. James, coolly denied any Catholic shame for his less than cunning subterfuge. All naked of divinity, a mortal creature with saline blood and sugar lumps beside the silver spoon. Carried on a throne through a world of wondrous believers, but for the insane who could not see his immortality fitted to their own peevish plans. Castrated in the bedrooms of the White House, his gonads lived in a mind befuddled by publicity and the sanctimony of the rich caring for the poor.

And Robert fought them in vain and fell, like Jesu, but in another battle, just as bloody, just as deadly.

/"This earth Pythonax and his brother hides, / Who died before they reached youth's lovely prime / The tomb their father built them; which abides / Forever, though they lived so short a time" /

Forgive me Father, for we have sinned.

And Fidel Castro sent Che Guerro further into the Peruvian mountains and to the jungles northward toward a Cabinet meeting's revenge. Begonias blossom like belladonna in the gardens of the Pentagon. The slide of the bolt action moves forward and backward to its work.

The click snapped and Charles knew another half hour had gone by. Five down; five to go.

Where, he wondered, were all the wild bulls from Arizona and Kansas, from California's movie world and Wall Street? Where were the money makers and genius destroyers when it comes time to die? How do the decision makers escape their destiny with truth?

L.B.J. and Subic Bay. Bourbon and branch water succeeding where Brown and Root would fail, to be overgrown and washed back to sea without a chance at those millions of gallons of stinking crude only 600 feet down off the end of the runway at Da Nang.

Sixty-eight miles of dead Americans, laid end to end, they'd stretch. Another 250 miles of wounded, end to end, the American way, and 250,000 men and women made sick by it all, sick beyond healing. And millions more, sicker with shame than they had ever been because the Gulf of Tomkin was said by Congress to be only nine thousand miles off shore and rising.

/"I hear, through dead men's drums, the riddled lads, / Strewing their bowels from a hill of bones, / Cry Eoi to the guns / My grave is watered by the crossing Jordan / The Arctic scut, the basin of the

South / Drip on my dead house garden. / Who seeks me landward, marking in my mouth / The straws of Asia, lose me as I turn / Through the Atlantic corn," /

Who can say, with any degree of certainty, which half of what is true? And Henry Kissinger told naught but lies.

The Mekong Delta ran south and its water was red with civilian blood as the enemy looked like friends. The rice was white when cooked, no matter what went in it or where it came from before it boiled. The merchants got fatter than the lice on the orphans' heads and their agents sold to agents who sold to agents, anything and all.

Soldier boys a-puking from drinking 3-Star kerosene and smoking hash in gangs, "too danged scared to die like men," the General said. The smoke hung heavy over every paddy and the night patrols were slaughtered because of their own stink and drunkenness.

So Harry Bruns got fragged because he was in a tent with a CID man the maligned blacks thought was an officer.

"One does not die often enough to really make it worth while. The reality of experiencing death should be repeated until you get it right," said Charles.

Needing a complete death, he was beginning to doubt it would be good enough the first time. His head was getting lighter and his circulation was poor. He felt the symptoms of a sniffling cold coming on. He got slowly to what had been solid feet and looked at the thermometer on the wall beside the door. It was 88 degrees.

"In the sun. Eighty-eight degrees in the sun," he said.

"I can't believe that makes any difference...what the hell does anything matter now?"

"Nothing..." he replied, smiling to himself.

He put another pill in his mouth and washed it down. His tongue was thick and his head became dizzy from standing. He was falling asleep. In a few moments he said, "What a hell of a way to go!"

His carbonated brain fizzled a stream of dreams that moved in and out of his fantasies. Few of them that would have awakened a toad.

Malignant Chile, his last adventure for the Company. Although he had once been safe behind a foreign border he had to return. The growing wonder and concern that such a thing as Allende could actually happen in North America's domain was resented. It was impure. It was, by God, un-American.

The guns and money came in and Dr. Allende was shot down without question, comment or mercy. Some witnesses said he was angry, furious at being killed out of context. Charles thought Allende should be angry. He was angry himself.

He reached for the table top to take the other tablets. His hand dropped before it reached them.

Dr. William Warren had mocked Charles Darwin Calvin

EPILOGUE

"The electoral triumph of the popular forces of my country has transformed into historical reality what up until now had been only stated as a theoretical possibility by classic Marxist thinkers. One that crowning day in September, 1970, Chile showed the world that the popular forces, when acting in the true spirit of struggle, admit no barriers, and that under the right conditions, the people using the legal and institutional arms created by the bourgeoisie, can seize the reins of government by electoral means to undertake the construction of socialism,"

"I have faith in my people. I also believe that in our path towards a more just society we can count on effective support from other people. Amongst these I include the North Americans, whose progressive tradition I respect; their new generations who generously strive for the creation of more just societies where all men can be truly free....."

Salvador Allende, July, 1971. Santiago,

Chile, S.A.

"Ants are gathered around an old tree.
In a choir they sing, in harsh and gravely voices,
Old Etruscan songs on tyranny.
Toads nearby clap their small hands, and join
The firefly songs, their five long toes trembling as
in the soaked earth"

...Robert Bly
Johnson's Cabinet Watched

by Ants